COOPERATIVE AGENTS

THEORY AND DECISION LIBRARY

General Editors: W. Leinfellner (*Vienna*) and G. Eberlein (*Munich*)

Series A: Philosophy and Methodology of the Social Sciences

Series B: Mathematical and Statistical Methods

Series C: Game Theory, Mathematical Programming and Operations Research

SERIES A: PHILOSOPHY AND METHODOLOGY OF THE SOCIAL SCIENCES

VOLUME 32

Scope: This series deals with the foundations, the general methodology and the criteria, goals and purpose of the social sciences. The emphasis in the Series A will be on well-argued, thoroughly analytical rather than advanced mathematical treatments. In this context, particular attention will be paid to game and decision theory and general philosophical topics from mathematics, psychology and economics, such as game theory, voting and welfare theory, with applications to political science, sociology, law and ethics.

The titles published in this series are listed at the end of this volume.

COOPERATIVE AGENTS

Applications in the Social Sciences

edited by

NICOLE J. SAAM
Ludwig-Maximilians-Universität München, Germany

and

BERND SCHMIDT
Universität Passau, Germany

SPRINGER-SCIENCE+BUSINESS MEDIA, B.V.

A C.I.P. Catalogue record for this book is available from the Library of Congress.

ISBN 978-90-481-5902-4 ISBN 978-94-017-1177-7 (eBook)
DOI 10.1007/978-94-017-1177-7

Printed on acid-free paper

Table of Contents

Hybrid forms of interaction between humans and artificial agents

Acknowledgements

It is a pleasure to express our thanks to Claudia Neumaier, who fulfilled the sometimes tedious task of preparing the author's files for the final camera-ready printout. The files came, as usual, in many different electronic formats. We also thank Markus Wagner, who did a qualified job in making up the list of concepts from sociology and computer science for the subject index. Many thanks to Edmund Chattoe who gave blood, sweat and tears to make this a proper English book in an extraordinary short period of time.

Finally, we thank the editors for including this volume in their series Theory and Decision Library.

München and Passau, August 2001

Nicole J. Saam Bernd Schmidt

INTRODUCTION

NICOLE J. SAAM
Institute of Sociology
Ludwig-Maximilians-Universität München
Munich, Germany

BERND SCHMIDT
Operations Research and System Theory
Universität Passau
Passau, Germany

Agents are a new paradigm which were first developed in computer science. In the meantime, agent technology has found its way into a variety of application areas. It has an especially important influence for the social sciences.

Interest in this field has been growing very rapidly world-wide. This fact prompts the necessity to give an overview about the efficiency and the restrictions of this new technology.

This book provides answers to the following questions:

- What are agents and how should this concept be used?
- What can agent technology contribute to the social sciences?
- What can sociology contribute to the development of agent technology?

From a sociological point of view, the essential features of human beings in a social context are autonomy, social ability, reactivity and pro-activeness. On the other hand, agents have properties like autonomous and intelligent behaviour, individual worldview and capabilities to co-operate and communicate. This comparison indicates that agent technology is a promising approach for the social sciences. At the same time, computer scientists who try to develop software systems bases on the concept of agents can profit from the knowledge of the social sciences about human action and interaction.

This book, therefore, presents contributions from both sociologists and computer scientists. It is the purpose of this book to demonstrate how agent technology influences the social sciences and how the social sciences in turn have an impact on computer science.

N.J. Saam and B. Schmidt (eds.)
Cooperative Agents. Applications in the Social Sciences, 1-4

The book is arranged in three sections which cover the following topics:

- Designing co-operative agents
- Agent-based social simulation
- Hybrid forms of interaction between humans and artificial agents.

Section one concentrates on designing co-operative agents. Both, computer scientists and sociologists give their views on this topic. Relevant questions are: What are agents and what are they for? What are agent models in software technology? The last contribution in this section is concerned with the question how social norms for co-operative agents could look like.

What are Agents and what are they for? *Bernd Schmidt* points out an essential difference between the use of this concept in social sciences, engineering sciences and formal theory. Subsequently, it is worked out in which way the agent technology is a useful methodology for modelling and simulating social systems. The explanatory value and the prognostic capacity are investigated and critically assessed. An architecture for the internal structure of an agent is proposed which is especially suited for modelling social systems. It is applied to role play among children.

From the point of view of software engineering, multi-agent systems are complex distributed application systems with a number of new properties, such as openness, dynamics, mobility and adaptivity. *Erika Horn* analyzes agents and agent systems from the point of view of software architecture. They are modelled in an object-oriented way, a pecularity being the explicit modelling of connectors as building blocks to couple agents. Examples of hierarchically structured sample architectures for classes of multi-agent systems and their relations to distributed artificial intelligence and sociology are discussed.

The importance of more adequate, non metaphorical social notions in Distributed Artificial Intelligence, such as norms, has long been emphasized. Coordination and common problem-solving among agents could require equivalents to social norms as they can be observed in human societies. *Nicole J. Saam* investigates into social norms for co-operative agents from a sociological point of view. She outlines how sociological theories define social norms and explain their emergence in human societies. She then suggests concepts of norms and norm processing that could either be used for planned implementation or for identifying equivalents of social norms among co-operative agents.

Section two is giving several examples of agent-based social simulation. These contributions are to show that agent technology really provides a new and useful methodology for research and application in the social sciences. Relevant questions in this respect are: Is agent based modelling really useful at all? How does fraud influence the degree of exchange and co-operation in a multi-agent system? How can the emergence of leadership from interaction in small groups be explained? Which facts determine social influence and how can these results be used to guide political actions?

Agent-based modelling offers a means of discovering general and applicable social theory, and grounding it in precise assumptions and derivations, whilst addressing those elements of individual cognition that are central to human society. However, there are important questions to be asked and difficulties to be overcome in achieving this potential. What differentiates agent-based modelling from traditional computer modelling? This and other questions are here addressed by *Jim Doran* and are illustrated by reference to recent agent-based models for the environment, e.g. the management of the Fraser River watershed in British Columbia.

A cognitive approach to cooperation and exchange allows to precisely individuate where and why the structures of exchange relationships have a different resistance to fraud and deception. *Rita Cecchi, Rosaria Conte* and *Cristiano Castelfranchi* provide an analysis of the cognitive plans that lead to a prosocial or a defective action, in order to demonstrate that the process of exchange has a natural bent to fraud regardless of external controls and detection, while cooperation is a naturally prosocial action. Their article also will demonstrate that a cognitive approach to cooperation and exchange is required to precisely individuate where and why the structures of these relationships have a different resistance to fraud and deception.

Nigel Gilbert and *Edmund Chattoe* explore whether social simulation research has to adopt a realist epistemology, or whether it can operate within a social constructionist framework. Research on leadership in small groups is used to illustrate the argument. The main question to be explored is how one can understand the emergence of identifiable leaders from interaction in small groups. The focus is on the emergence of 'macro' phenomena (in this case, leadership and follower roles from an undifferentiated collection of agents) as a result of interactions between agents at the 'micro' level.

In the field of social psychology, there are several theories that specify the variables and processes of human interaction. These theories have found little representation in the literature on simulation, however. In response to this deficit, *Hans-Joachim Mosler* and *Wernher Brucks* discuss simulations that have made social influence a central topic. One of their own projects, a simulation of the theory of social comparison, explicates the challenges and findings of simulations grounded in social psychology.

Section three concludes with hybrid forms of interaction between humans and artificial agents. It shows the close relation between agent technology in the social sciences and in computer sciences. Whereas sociologists of technology analyse hybrid organizations, computer scientists develop agent-based learning systems that create hybrid learning communities or agent-based systems for hospital management.

The more traditional approaches to implement information systems in organizations often lead either to dysfunctional restrictions imposed on the organization by the information system's need to establish firm cause/effect relationships. Or, reversely, they lead to incoherence imposed on the information system by the organization's structure. *Ingo Schulz-Schaeffer* describes the theoretical background of the INKA research project that aims at developing a solution to the problem sketched above by modeling and exploring hybrid forms of interaction between humans and artificial agents within organizational settings.

Learning by help of tutoring systems is a growing market. Intelligent distributed learning environments are applications in which hybrid communities of human actors and artificial agents evolve. *Andreas Harrer* presents the design and features of intelligent distributed learning environments and actors that participate in hybrid learning communities. The learning system is designed as a multi-agent system of cooperative intelligent agents. Human learners may take different roles in the learning process.

Traditional modeling approaches mainly stemming from technically oriented application domains more and more turn out to fail in supporting the economic and organizational planning process in the hospital domain adequately. One of the major reasons for this problem arises from ignoring human decision making and behaviour as a relevant influence on the performance of such systems. *Rainer Sibbel* and *Christoph Urban* show how this problem could be overcome by integrating agent technology into current modeling approaches.

This book is intended to give a concise overview of the state of the art in agent technology. It shows the importance and the promising prospects for sociology and computer science alike. The editors hope that this book supports and promotes future research. At the same time the editors mean to give new ideas on possible applications of this new and fascinating field.

WHAT ARE AGENTS AND WHAT ARE THEY FOR?

BERND SCHMIDT
Operations Research and System Theory
Universität Passau
Passau, Germany

Abstract

Agents are a new modelling paradigm which has an important influence for the social sciences. First a definition of the concept *Agent* is given and the application areas are described. An essential difference arises between the use of this concept in social sciences, engineering sciences and formal theory.

Subsequently, it is worked out in which way the agent technology is a useful methodology for modelling and simulating social systems. The explanatory value and the prognostic capacity are investigated and critically assessed.

The third part proposes an architecture for the internal structure of an agent which is especially suited for modelling social systems. It turns out that that agents in the social sciences should have the descriptive capacity for emotional reactions and for cognitive processes as well.

An detailed example for the application of agents in the social sciences is given. It investigates the role play among children.

Agent-based systems are of increasing importance. They are regarded as a new paradigm enabling an important step forward in empirical sciences, technology and theory. They are of particular importance in the social sciences.

1. The concept *Agent*

This new paradigm has an enormous influence on model building and simulation. Novel design patterns and architectural structures are opening up new possibilities. Furthermore, new areas of application are coming into the foreground.

1.1. ATTRIBUTES OF AN AGENT

Cooperative agents can be equipped with different attributes which however do not all appear to be necessary or useful in every application. The following are

N.J. Saam and B. Schmidt (eds.)
Cooperative Agents. Applications in the Social Sciences, 5-20.

of particular importance:

- **Autonomous Behaviour**. Every agent is characterised by autonomous behaviour. This consists in the first instance of the so-called inherent dynamism that an intelligent agent displays autonomously without input from the environment.

  In addition there is the induced dynamism which describes how the intelligent agent reacts in response to inputs from the environment.

  Strategies that guide the behaviour of an agent when it is pursuing its goals are of particular importance.

- **Individual World View**. Every agent has its own model of the external world that surrounds it. This so-called conceptual model describes how the intelligent agent sees the world. The conceptual model will generally be incomplete. In some circumstances it is even incorrect.

  The manner in which the intelligent agent builds up its model of the world on the basis of the information it receives from its environment is of particular interest.

- **Communicative and Cooperative Capacity**. Intelligent agents can exchange information with their environment and with other intelligent agents. By means of the possibility of communication an intelligent agent must obtain information about its environment which enables it to build up its own world model. Moreover the possibility of communication with other intelligent agents is the precondition of common action in pursuit of a goal.

- **Intelligent Behaviour**. As the term *Intelligent Agent* indicates, behavioural possibilities such as the capacity to learn, logical deduction or construction of an environmental model in order to find one's bearings in an unknown environment and suchlike are required.

- **Spatial mobility**. Intelligent agents are sometimes required to display spatial mobility. Spatial mobility is in fact a characteristic of numerous intelligent agents. On the other hand there are elements in the real world that are not spatially mobile but which one would nonetheless wish to assign to intelligent agents. For this reason spatial mobility in this study will not be considered as one of the necessary attributes of intelligent agents.

The modelling and simulation of real systems consisting of intelligent agents that cooperate with each other has recently emerged as an important field of research. Systems of this kind are of particular importance in the social sciences. In this context two areas are of particular interest:

- **Strategies and Decentral Control**. The object is to develop individual strategies that individual agents pursue and that ensure that a common goal can be achieved even without central regulation.

- **Emergent Behaviour**. Cooperation between intelligent agents can produce a stable system that displays new global behaviour on the next higher level of abstraction. The task is to explain this global behaviour on the basis of the individual attributes of the intelligent agents and of their interactions.

1.2. WHAT ARE AGENTS AND WHAT ARE THEY FOR?

Agents have been used with convincing results in science, technology and in theory. It is useful to distinguish between the deployment of agents in these three fields in order to avoid misunderstandings and to reduce unreasonable expectations.

Social sciences as all empirical sciences investigate real systems and develop models for them. This means that social sciences find their objects of cognition and have to take into account the conditions that they encounter in the real world. System theory has shown that numerous real systems despite superficial differences have the same deep structure. For all models which resemble each other in this way uniform basic structures of models can be designed. Such models, which can be applied to a clearly-defined class of real systems, will be referred to as reference models (Urban 2000).

Once reference models are developed on the basis of agents, the internal structure of the agents must conform to the objects in reality which are to be depicted by means of agents. The qualities and modes of behaviour with which an agent is endowed are determined by the qualities and the behaviour of the real objects which are modelled by means of agents.

This is also true of real systems in which human beings are involved and for which agents are deployed as models for real human beings. The agents acting as representatives for real human beings in the model must be as similar as possible to real human beings. They should possess similar qualities and display similar types of behaviour.

It will become clear that human behaviour does not consist solely of cognitive components but is to a significant extent emotionally conditioned. It therefore follows that agents whose function is to simulate human behaviour must natural also possess emotional capacities (Moffat 1997).

We can see that in order reliably and conscientiously to model human beings and human behaviour we cannot avoid posing the question of the nature of human beings.

1.2.1. Agents in Engineering Sciences and in Technology

Engineering sciences and technology have the task of designing and constructing new systems, systems which have not previously existed. In contrast to the empirical sciences, engineering sciences and technology are not primarily concerned with accuracy of depiction and structural resemblance to an already existing system. Here the emphasis is on function and performance. When developing and implementing new technological systems the engineer has to meet the following requirements: to produce a model which is clear, simple, functional and reasonably cheap. Apart from these requirements, no limits are imposed on his inventiveness and his creativity.

The engineer confronted with such tasks also has standard solutions at his disposal: they know for example how a suspension bridge or a tunnel is to be built. These standard solutions will be referred to as architecture patterns.

It has proved useful to have recourse to architecture patterns based on agents for certain tasks. One thinks in this context of the design of robots or of complex software systems. When agents are deployed in this context, their internal structure and internal architecture conform only to the requirements of engineering science. It is homo faber who operates in this field.

This means that the question of whether robots or software agents should be provided with emotions is solely a question of whether such agents function better and can be more cheaply produced than a design with a differently structured architecture pattern.

1.2.2. Theory and Virtual Worlds

The human mind is free and cannot be constrained by the instructions of nature for which a model is to be produced or by the requirements of a technical device which is to be constructed. It is capable of thinking up new realities and of trying out new possibilities. This does not mean that the virtual world created in this way must immediately be regarded as a model for actual real systems or as a blueprint for technical systems. Human beings develop systems with arbitrary qualities in a playful, intention-free manner and curiously study the modes of behaviour these new systems display. We are dealing here with homo ludens.

Nobody would dream of criticising such systems because they did not replicate reality or were useless and without practical application. If a system that originated in the virtual world contains agents, these agents can be however their designer wishes them to be. They can contain emotions or no emotions, as he wishes. The entertainment and science fiction branches make liberal use of this possibility. Criteria for the meaningfulness of such activity can be originality, creativity, imaginativeness and entertainment value.

1.2.3. Synergies

So far we have stressed the fact that the deployment of agents is meaningful in the social sciences, in technology and in the creation of virtual worlds but that in all three cases different types of demands are made on the agents. In this context it seems useful to sum up the essential points:

This could create the impression that the three areas of deployment are completely watertight and have nothing to do with each other. This is not the case. There are interactions and reciprocal influences. For example social sciences frequently find inspiration as blueprints for reference models in constructs of theory which at first seemed to have no application. Engineering sciences try to adopt construction schemata from models of nature developed by social sciences and then try to use them as the basis for architecture patterns. Socionics and bionics are examples of this. Theory in turns takes up conceptions from the social sciences or from technology and develops them further.

It is therefore advisable, whatever field one is working in, to risk taking a look over the garden fence in order to find inspiration and enrichment of one's ideas.

Field	Result	Requirement
Social science	Reference models	Structural similarity
Engineering science	Architecture patterns	Functionality
Theory	Virtual worlds	Originality

Diagram 1: Fields of application for agents

2. The Modelling of Human Behaviour in the Social Sciences

The first question is whether the attempt to investigate human beings scientifically and to capture their nature in a model is at all possible and whether it is not an example of hubris.

2.1. THE UNFATHOMABLE NATURE OF HUMAN BEINGS

Human behaviour is determined by a wide variety of influencing factors which also interact in complex ways with each other. A few examples chosen at random will serve to illustrate this:

- **Personal Experience**. The present behaviour of a human being is undoubtedly influenced by his life history and by the experiences he has had. These include early childhood experiences as well as consciously learned experience about the environment.
- **Social Norms and Role Expectations**. Every human being to a certain extent conforms to the norms and role expectations which society imposes on him. These demands often conflict with his individual wishes, plans and ambitions.
- **The Unconscious**. Human beings are very often not conscious of their

own actions. Again and again they act or react in ways which surprise them or which they did not expect of themselves. Is not St Paul right when he says: „*I do not do the good that I want to do, but I do the evil that I do not want to do?*“ (Rom. 7, 19).

- **Conflicting Motives**. In conflict situations human beings are torn between different motives that are at variance and conflict with each other. Should they in the Kantian sense for example do their duty or should they rather follow their inclinations?
- **Human Freedom**. One also has to take into account free will, which enables human beings themselves to decide what to do, who to trust, what roles to play or what norms to follow, quite independently of external influences and stimuli and solely on the basis of their own personality and their own free decision.
- **The Experience and Response to Art**. In the arts human beings have access to a world of experience which eludes rational access. Beethoven noted in his diary: „*Music is a higher revelation than all wisdom and philosophy.*“

Can all this be modelled?

Can we really fully understand human beings. Will we really be able to reproduce their behaviour in a computer model?

Is the human being incomprehensible?

Of course it can be assumed that human behaviour is very complex and many-layered. The present research project however is based on the conviction that it is possible to reduce this complexity by means of conscientious functional decomposition and to dissolve this multi-layered quality by carefully isolating the individual layers. These layers are first studied in isolation and then their interaction and their interplay are investigated.

It is hoped that by this means we can approximate to a deeper understanding of human behaviour and make human behaviour comprehensible.

2.2. THE DIFFERENCE BETWEEN MODEL AND REPLICA

In order to try to answer the question whether human beings can be really modelled by agents, we must first point out the difference between a model and a replica.

A replica is an identical copy of an original. A replica is completely indistinguishable from the original. It appears to be impossible, at least for the foreseeable future, to produce a replica of this kind.

A model on the other hand is an abbreviated depiction of an excerpt of reality based on abstraction and idealisation. It does not have to conform to reality in every aspect and in all respects. The example of the aeroplane in the wind tunnel illustrates this point. Such insights are valid in the human sciences too. For example literary and historical scholarship have developed a picture and hence a model of Goethe. This picture is of course not identical to the real Goethe. It does not claim to be a replica. Nonetheless it does provide useful and useable insights. The more precise and the more detailed our image and thus our model of Goethe is, the better he can be understood and the more accurately it will be possible to predict his behaviour in a certain situation. Not exactly of course, but in terms of a general tendency. If we had a good model of Goethe, we would know something about his physical condition, his emotional state, the state of his knowledge of the world and his social position. It would then be conceivable that we understand for example why at an advanced age he falls in love with a young woman. We would even perhaps be in a position to forecast something of the kind. We would not be able to state exactly where and when this would take place. But we could assume that it would happen.

A good, useable and useful model capable of providing valuable insights does not necessarily have to be insurmountably complex and difficult. It could turn out in fact to be quite simple, so simple that modelling could be successful. This means that a model of a human being does not necessarily have to contain all the qualities that distinguish the human being as human being. One can try to begin in a simple way and to concentrate on the dominant facts in the case of the problem under investigation.

Applying this to the children's play group, this insight leads us to hope that such a project is not impossible from the outset.

In order to understand the play behaviour of children and to make it predictable within limits, it is not necessary to model the children in all their complexity. There is no need for a replica. Many qualities and modes of behaviour that normally typify a child can be sacrificed to the filtering abstraction and idealisation process without this rendering the modelling process completely futile.

2.3. MODELS OF HUMAN BEHAVIOUR IN THE SOCIAL SCIENCES

When considering the modelling of human behaviour, it should be borne in mind that excellent models of the human being and his behaviour already exist for certain areas.

Physiology has developed very detailed and expressive models of the human physis and of its behaviour in changing circumstances. It is possible to model, to understand and to predict physical and chemical processes in the human body.

In a similar manner psychology attempts to develop models of human psychological life which deepen our understanding of internal processes. This category includes cognitive aspects such as intelligence, learning, memory and powers of imagination. In addition, considerable work is being done to improve our understanding of emotional dynamism in its healthy as well as in its pathological forms.

Sociology also works with models and attempts to understand human behaviour in its non-individual form, i.e. in relation to society. Sociological models investigate for example the development, the passing on and the implementation of norms or the development of norms. In both cases sociology attempts to identify human behaviour in social groupings.

If one were to question the explanatory value and the prognostic capacity of these models, then physiology, psychology and sociology would cease to exist. No one would wish to adopt such a stance.

Everyday experience provides a further example. The better one knows a person, the better one is generally able to understand this person and in some circumstances even to predict his behaviour.

These arguments have shown that it is perfectly possible at least in certain areas to model and to understand human behaviour and to make it predictable. The important proviso remains that for human beings too a model is always an abbreviated and crude version of the original and can never be identical to it. A model of the human being is fundamentally different from the human being himself. Nevertheless a model is useful and meaningful.

Many critics who doubt the possibility of modelling the human being are not aware of this point. They confuse a model with a replica and because of this become engaged in incomprehensible polemics.

2.4. THE HUMAN BEING AS A PSYCHOSOMATIC UNIT WITH COGNITIVE ABILITIES IN A SOCIAL ENVIRONMENT.

The sciences have so far concentrated on investigating partial aspects of human behaviour under the laboratory conditions appropriate in each case. As a result there has been a tendency to lose sight of the interactions between the emotional, the cognitive and the social area. It is a fundamental conviction of this research program that an understanding of human behaviour can be achieved only if all 4 aspects and their interaction are taken into account. According to this interpretation, the human being is a psychosomatic unit with cognitive capacities who is capable of surviving only in society. His behaviour will always be determined and shaped by the interaction between his physical situation, his emotional state, his cognitive capacities and his social position.

As soon as one attempts to model real human behaviour, it is essential to have a reference model that permits the possibility of this interaction and this interplay. Engineering sciences with their architecture models and theory with its virtual realities do not need to take this possibility into account.

2.5. BEHAVIOUR CONTROL

Generally speaking we can start by assuming that every organism has certain needs which it wishes to satisfy. In the course of time, evolution has constantly developed improved and more powerful forms of behaviour control in order to guarantee the satisfaction of these needs. For reasons of clarity it is useful to distinguish the following forms of behaviour control:

- Reactive behaviour
- Deliberative behaviour

We can assume that the human being as a product of evolution has all these modes of behaviour control at his disposal. He is capable of the highest form of reflective behaviour without having completely liberated himself from elementary forms of instinctive behaviour. The human being is a citizen of several worlds.

A reference model that presumes to model human beings as a whole must provide an architecture that makes it possible in principle to model all forms of behaviour control.

3. Agent Based Models for Behaviour

The various forms of behaviour each have different preconditions that have to be met before the corresponding form of behaviour is capable of functioning. For example learned behaviour requires the ability to absorb and to process information. These abilities would not be necessary for instinctive behaviour.

In order to develop reference models for behaviour, we must first consider how we can picture the mode of functioning of the various forms of behaviour, what the preconditions for these forms of behaviour are and how they can be depicted in a model.

An initial classification distinguishes between reactive and deliberative behaviour. Reactive behaviour operates according to fixed, pre-set rules. Here the focus is on the set of regulations on the basis of which the various actions are brought about. Deliberative action in contrast pursues a goal by means of conscious reflection, by working with models and by trying and testing in the mind's eye. The constitutive element here is the plan of action that must be drawn up in order to achieve the goal.

3.1. REACTIVE BEHAVIOUR

With reactive behaviour there exists a set of rules which can be called up when the need arises and which provide a course of action appropriate to the circumstances. No thought processes are required or possible.

Reactive behaviour can be subdivided into the following categories:

1. Instinctive behaviour
2. Learned behaviour
3. Drive-controlled behaviour
4. Emotionally driven behaviour

3.2. DELIBERATIVE BEHAVIOUR

What the reactive forms of behaviour described so far have in common is that the goal or the purpose that the actions are intended to serve are not known or not conscious. An organism in this sense does not „know“ that its actions serve the purpose of need satisfaction and thus survival. Deliberative behaviour, on the other hand, involves the conscious pursuit of goals. The organism can on its own initiative devise the most favourable action plan for achieving its goal.

This action plan contains partial goals that are necessary to attain the ultimate goal. Each individual partial goal consists of a sequence of actions in which individual actions are reactively performed on the basis of rules.

In order to draw up an action plan and the sequence of actions necessary for each partial goal, an optimisation plan will as a rule be involved.

We have divided deliberative behaviour into two types:

- **Constructive Behaviour**. The goal is known but cannot be changed. An action plan is consciously constructed that leads to this goal.
- **Reflective Behaviour**. The organism has the possibility of self-regulation. It can set its own goals or it can modify existing goals.

Deliberative behaviour is an extension which considerably increases scope for action in comparison with reactive behaviour. It also implies improved adaptation to the environment, a more efficient satisfaction of needs and hence increased chances of survival.

3.3. MOTIVES AND MOTIVE SELECTION

Motive is defined as a psychological force that drives the organism to execute an action. Motives can be:

- Drives
- Emotions
- Acts of will

What all motives have in common is that they appear with a certain intensity and that they may also be in competition with each other. The motive with the highest motive intensity will determine action.

As drives, emotions and acts of will together are regarded as motives and as each of these motives has a corresponding intensity, these motives are comparable. It is possible to establish which motive is strongest at a given time and so determines the action to be executed. It is conceivable that an agent experiences hunger and rage simultaneously while also following the goal of tidying the house. We are then confronted with the following state of affairs:

- Drive: hunger

  Selection criterion: intensity of drive

  Drive-controlled behaviour: Go to the fridge

- Emotion: rage

  Selection criterion: intensity of emotion

  Emotionally controlled behaviour: Kick the door in

- Act of will: tidy the house

  Selection criterion: strength of will

  Deliberative behaviour: tidy the room

The three different selection criteria that determine action, namely drive intensity, emotional intensity and intensity of will, are not constant but change with time. This means that different motives may be action-determining at different times. Thus it is possible that at first the act of will *Tidy the house* displays the highest intensity of motive. However after a while hunger becomes stronger and stronger. It will then overtake the intensity of will that led to the action of tidying. The action of tidying is interrupted. A new motive takes over control. The agent will go the fridge.

For the modelling of human behaviour there will have to be a motive selector that monitors the development of various motive intensities and ensures that it is always the strongest motive that determines action.

Diagram 2 shows the connection.

On the left side are facts which could be the causes of motives.

- **Energy**. Every organism needs energy. An energy deficit will result in hunger as a drive. The lower the energy, i.e. the higher the energy deficit, the greater the hunger will be.
- **Failure**. The fact that one experiences failure leads to the motive of anger. This anger operates as a motive whose motive intensity can be calculated. An unimportant setback will result in a weaker emotion than the failure of an important project.
- **Knowledge**. An act of will is presupposed that has consciously set itself the goal of acquiring knowledge about an unknown environment. The

will intensity that operates as a motive can be determined. The importance of the goal is also taken into account here.

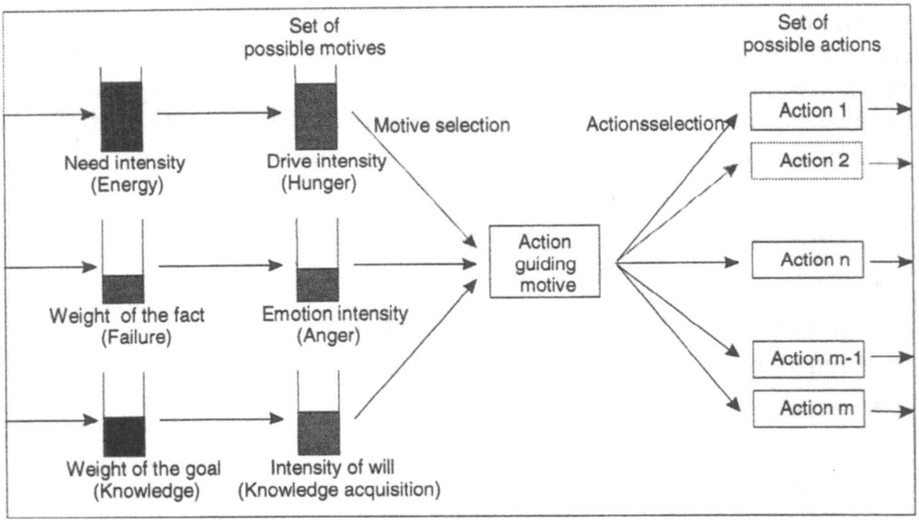

Diagram 2: Motives and motive selection

The motive intensities of hunger as a drive, anger as an emotion and knowledge acquisition as an act of will are in competition with each other. The motive with the highest motive intensity is chosen as the action-guiding motive via motive intensity. From the set of possible actions it determines the action or action sequence that is actually carried out. In general the execution of the action will impact on the original state and modify it.

In *diagram 2* hunger has the highest motive intensity. It will determine the action to be executed. This could for example be food search and food intake. These actions will improve the energy level and thus reduce hunger. It follows from this that instead of hunger for example the act of will of knowledge acquisition will become the strongest and thus the action-guiding motive. The action of exploration of the environment could then come into play.

The central idea of motive intensity and of the motive selector has been taken from (Dörner 1999). It was developed and adapted to the basic concepts of the PECS reference model (Schmidt 2000) and (Urban 2000).

4. The Role Play Model

The role play model belongs in the area of social psychology. In this model a real group is studied and modelled. Its members are children who perform various role plays, for example shopping, mothers and fathers or cowboys and Indians. The processes of group formation, of group activities and of group disbandment are of particular interest here.

A more detailed description of the Role Play model will be found in (Ohler and Reger 1999) and (Mogel 1994).

Role play among children is particularly suitable for the study of group behaviour. Emotional reactions by children are easily observable, because they are hardly distorted at all by social control mechanisms or by learned modes of behaviour. Moreover the results of extensive experiments in this field are available (Mogel 1994).

The object of the model is theory-validation. First the play behaviour of the children is analysed and described on a superordinate level. These empirically collected data are to be understood, explained and to a certain extent made predictable. They are interpreted as emergent behaviour that results from the communication and cooperation of the children at the next lowest level of observation.

4.1. THEORY VALIDATION BY MEANS OF MODELS

In human sciences such as psychology or sociology for example theory validation is difficult because human beings as the objects of study cannot be subjected to various experiments as in the natural sciences. Here system simulation opens up new opportunities. It makes possible the construction of a model of a real system that is studied and for which a theory is to be formulated and validated. Hence it is no longer necessary to carry out the proposed experiments on real human beings. Instead a model can be constructed in accordance with the theory. The resulting consequences can then be studied in the model and the model results can be compared with the observed physical data. If the correspondence between the empirically collected system data and the simulated model data is satisfactory, then an important step towards theory validation has been taken.

The task is to develop a theory for the play behaviour of children during role play. The object is to understand and to explain why children under certain external conditions behave in a certain way towards each other.

In order to do this it is first necessary to analyse and to conceptualise play behaviour on the higher level of abstraction. It has to be empirically established which play forms are preferred, how long on average a role play may last, in which circumstances existing play groups disband and new play groups form and so on.

A theory that wishes to explain these empirically collected and described conditions must see them in terms of emergent behaviour, which occurs when individual children with their respective qualities and attitudes come into contact with each other.

The first task of the theory is experimentally to identify those qualities, attitudes and modes of behaviour which are necessary in order to explain the observed behaviour on the superordinate level. From the beginning it is not at all clear which qualities, attitudes and modes of behaviour from the variety of possibilities are actually relevant and significantly influence the children's play behaviour as opposed to merely producing small, unimportant side effects. The second task of the theory is to link these qualities, attitudes and modes of behaviour in such a way that the play behaviour of the children in the group results.

4.2. QUALITIES, ATTITUDES AND MODES OF BEHAVIOUR

Every attempt to comprehend a field of study begins with a certain preliminary comprehension which is at first vague. Certain assumptions are made with which the object of study is approached.

In the case of a theory for children's role play, years of experience have shown that it is sensible to consider the following facts:

- **Children's individual qualities**. These are constant qualities which constitute the child's personality and which hardly change in the course of the game. These may include for example:

  1. Aggressiveness or calmness.
  2. Extroversion or introversion
  3. Curiosity or lack of interest
  4. Linguistic competence

  These qualities are treated as constants in the model.

  In addition there are the variable qualities which can change in the course of the game and are dependent on the course of the game. These may include for example:

  1. Emotions
  2. Pleasure in the game
  3. Satisfaction with the game

  These changing qualities are treated as state variables in the model.

- The attitude of the children to other children and to the game. (For the concept of attitude see Krech and Crutchfield 1992).

  The attitude of an individual child to other children and to the game includes the following elements:

  1. Factual knowledge
  2. Evaluation
  3. Emotional attitude
  4. Arrangements for action

In the case of attitudes it is also sensible to distinguish between constant attitudes, which change only slightly in the course of the game, and attitudes which are subject to rapid change during the game.

4.3. VALIDATION

In order either to verify or to falsify the validity of the abstract model and thus of the theoretical approach, the model data are compared with the empirically collected system data.

Detailed empirical data have to be collected for the scientific study of children in role plays. The following methods are available:

- **Individual experiments in the laboratory**. Individual experiments are conducted in the laboratory. The aim here is to gather information about the child independently of the play environment. Such experiments would include for example personality diagnostic tests for example to determine aggressivity, stamina etc. Preference for game types and for roles as well as exploratory behaviour can also be identified in this way.
- **Data collection in the field**. Important information about the specific responses of children can best be collected in the environment familiar to the child, i.e. the kindergarten. Such specific responses include styles of interaction, communicative competence, social cognition etc. (Mogel 1994).

  In addition to this, partly structured interviews take place in the domestic settings. Here the most important eco-system information for the domestic milieu is gathered, for example the attitudes of the parents to education, to cooperation processes and to the child's play behaviour.
- **Group experiments in the laboratory**. A powerful research laboratory is available in which the group behaviour of children can be studied in detail. Here relevant information for play behaviour and thus for model validation is collected.

5. Conclusion

Agent based modelling opens up a new possibility of theory validation in human sciences, especially in sociology. The consequences of theoretical assumptions can be investigated. The comparison of system and model data provides an indication of the validity of the theory or of the abstract model used. The role play model shows the method to be used.

References

Dörner, D. (1999). *Bauplan für eine Seele*. Rowohlt Verlag, Reinbek bei Hamburg.

Goleman, D. (1996). *Emotional Intelligence*. Bloomsbury Publishing Plc, London.

Krech, D. and Crutchfield, R.S. et al. (1992). Einstellungen und Einstellungsänderungen. In: *Grundlagen der Psychologie*, Band 7 Sozialpsychologie. Beltz Psychologie-Verlags Union, Weinheim.

Moffat, D. and Frijda, N.H. (1994). Where there´s a Will, there´s an Agent. In: Wooldridge M.J. and Jennings, N.R. (eds.). Proceedings of the 1994 Workshop On Agent Theories, Architectures and Languages (pp. 245 – 260); Springer Verlag, New York.

Moffat, D. (1997). Personality Parameters and Programs. In: Trappl, R. and Petta, P. (eds), *Creating Personalities for Synthetic Actors* (pp. 120-165). Springer Verlag, New York.

Mogel, H. (1994); *Psychologie des Kinderspiels*. Springer Verlag, Berlin.

Ohler, P. and Reger, K. (1999). Emotional Cooperating Agents and Group Formation - A System Analysis of Role-Play among Children. In: Szczerbicka, H. et al. (eds.): *Modelling and Simulation - A Tool for the Next Millenium*. Proceedings of the 13th European Simulation Multiconference, Warsaw, Poland, June 1999, SCS Publications; San Diego.

Schmidt, B. (2000). *Die Modellierung menschlichen Verhaltens*. SCS-Verlag, Erlangen.

Urban, C. (2000). PECS - A Reference Model for the Simulation of Multi-Agent Systems. In: Suleiman, R., Troitzsch, K.G., and Gilbert, G.N. (eds.). *Tools and Techniques for Social Science Simulation*. Physica-Verlag, Heidelberg.

AGENT MODELS OF SOFTWARE TECHNOLOGY[1]

ERIKA HORN
Chair of Software Engineering
Institute of Computer Science
University of Potsdam
Potsdam, Germany

Abstract

From the point of view of software engineering, multi-agent systems are complex distributed application systems with a number of new properties, such as openness, dynamics, mobility and adaptivity. In this article, agents and agent systems are analysed from the point of view of software architecture. They are modelled in an object-oriented way, a novelty being the explicit modelling of connectors as building blocks to couple agents. Hierarchically structured sample architectures for classes of multi-agent systems can be given by the abstraction of existing agent applications and by using object-oriented principles. The article provides examples of sample architectures and their relations with distributed artificial intelligence and sociology. The advantages and disadvantages of agent technology are also discussed.

1. Multi-agent systems and software engineering

In recent years, there has been an increasing focus on multi-agent systems by scientists and users in various fields. This focus has generated extensive theoretical work as well as applications of agent technology in real-life situations. Although the original roots of multi-agent systems lie in the field of distributed artificial intelligence (DAI), interdisciplinary activities are on the increase. The view of agents and agent systems varies across the individual disciplines of computer science. This article focuses on views that have been influenced by the FIPA (FIPA 1998) and MASIF (Mobile Agent Systems Interoperability Specification 1997) standards. It looks at agents and agent systems from the point of view of software engineering.

A ***software agent*** is a computer program working autonomously and with goal-orientation in the name of a person or an organisation (authority). An agent has got its own execution profile and is able to execute tasks on its own

[1] This article is the written version of a talk given during the workshop "Multi-agent Systems" held in Tutzingen in June 1999.

N.J. Saam and B. Schmidt (eds.)
Cooperative Agents. Applications in the Social Sciences, 21-37.

initiative and cooperating with other agents during its life cycle. An agent has a property determining who is authorised to use it. Mobility can also be an essential property of agents.

From the software engineering perspective, a multi-agent system is a complex distributed application system consisting of building blocks. These building blocks in this case are the agents and their interconnections. Central aspects from the perspective of software engineering include:

- openness, dynamics and adaptivity of multi-agent systems
- building blocks for multi-agent systems and building block libraries for agents and their connections, reuse of building blocks, patterns, subsystems and application frames
- multi-agent systems as complex and heterogeneous software systems, e.g. integrating existing applications and system services with new ones
- application specific security aspects in multi-agent systems
- software ergonomic aspects of multi-agent systems
- workflow modelling and role orientation
- methods and tools for the development of multi-agent systems
- methods (and criteria) for quality assessment of multi-agent systems

Agent technology is based to a considerable extent on the ideas of object-oriented software development. In the multi-agent case, implementation of the above aspects in software technology is more or less completely absent. Newer activities in standardising agent technologies such as FIPA (FIPA 1998) and OMG (Mobile Agent Systems 1998) only take into consideration selected aspects, e.g. agent management, communication, mobility or interoperability. Methodological questions are not treated in the standards.

Based on the notions of software architecture and architecture type (section 2), models of agents and agent systems using object-oriented principles will be discussed in this article. Object-oriented modelling offers a number of benefits when analysing and constructing agent systems. It is an excellent basis for the prefabrication and re-use of software building blocks for example. On the basis of object-oriented models, common sample architectures for multi-agent systems and architectures specific to an application domain can be defined (section 3). Agent application systems are developed "into" these sample architectures. Thus the sample architectures constitute knowledge about classes of agent application systems. Agent technology using sample architectures thus realises an engineering-technical principle in the construction of systems that has long been commonplace in other engineering disciplines, such as mechanical or electrical engineering.

The time has come to unify the various views, scientific approaches and practical experiences of the individual disciplines of computer science and related sciences into a science of multi-agent systems. That is why section 4 deals with some questions of agent technology (from the point of view of software engineering) which are related to DAI and sociology.

Finally, section 5 contains a discussion about the advantages and disadvantages of agent applications and an overview from the perspective of software engineering.

2. Architecture models for multi-agent systems

The object-oriented paradigm is used to analyse agent application systems from the point of view of software architecture. This section introduces basic notions of software architecture and their interpretation for agent applications.

2.1. FROM ARCHITECTURE SAMPLES TO ARCHITECTURE TYPES

Conventional architecture notions mostly refer to the structural properties of software systems. They describe the properties on a high level of abstraction avoiding implementation details (System Application Architecture 1998, Reference Model for Frameworks 1993). An operational view, supplementing this structural view, is discussed below. This allows us to implement the methods for construction of software systems.

Referring to Shaw and Garlan (1996), we begin by introducing notions of software architecture and architecture pattern (architecture style). The software architecture of a software system comprises a description of the elements which the system contains (components) and the coupling (and interaction between) the elements (connectors). For classes of software systems, architecture frameworks can be given. These frameworks define the types (syntax and semantics) of the elements and interactions as well as the coupling rules for those elements and interactions. Furthermore, tests to establish whether a software system meets the requirements of the architecture pattern can be produced. Thus architecture frameworks are the abstracted knowledge of the software developer. Examples of well-known architecture frameworks are pipe filter systems, layered systems, blackboard systems, repositories and so on.

In order to take into consideration the operational view of software architectures too, the following additional notions are introduced. Architecture operations are related to an architecture pattern and describe the generation, manipulation, transformation and analysis of architecture specimens. An architecture type is an architecture pattern with the related architecture operations. An architecture specimen is a realisation (instance) of an architecture type.

As discussed in Horn, Kupries and Reinke (1999), the object-oriented paradigm is particularly suitable to describe architecture types in terms of their structural and operational properties. Abstractions, more detailed descriptions, compositions and analyses are made possible by using object-oriented concepts. Architecture types can be created from a common basic architecture type by derivation and interpretation and can thus be structured hierarchically themselves. Within the hierarchy, the applicability of the architectures and their building blocks increases.

The object-oriented notion of architecture introduced here is very suitable to model multi-agent systems for the following reasons:

- It allows the explicit modelling of components and connectors. The latter are of particular importance in agent systems with their varied relations between agents
- It promotes the abstraction of reusable building blocks and building block libraries as well as the use of object-oriented concepts for reuse (such as instancing, inheritance and composition)
- It allows an integrated overall look at the structure of an agent application and at the operation of its construction
- Using the object-oriented paradigm for all phases in the construction of agent systems achieves methodological coherence

2.2. OBJECT-ORIENTED ARCHITECTURE MODELS FOR MULTI-AGENT SYSTEMS

By interpreting object-oriented architecture models, a special basis architecture type for agent applications can be developed. It contains agents and agent connectors as building blocks as well as analysing their composition into agent systems.

Since **agents** are the execution units in an agent application, they are modelled as special components.

Agent connectors provide the communication and interaction between agents in an agent system. Agent connectors provide a protocol regulating the interactions between agents, enabling one agent to use the functionality of others or allowing it access to external resources. Because of the agents' autonomy, goal-orientation and great demand for interaction, the explicit modelling of connectors as building blocks is of particular importance in agent systems. This is explained with the help of an example in section 3.1.

An **agent system** contains agents and agent connectors and manages the relations between them. It provides operations that allow the generation, interpretation, execution and termination of agents and enables the establishment of connections (connectors) between agents and authorities and between the agents themselves. In the object-oriented sense, agent systems are instances of a composite class. The left hand side of *figure 1* illustrates relations between the building blocks of the basis architecture type for agent applications. These are depicted by means of expressions in the Unified Modelling Language (Rumbaugh, Jacobson and Booch 1999). The model contains the classes for CAgent, CAgentConnector and CAgentSystem, classes for the properties of the basic components as well as root classes for the building block and property classes. Class names have the prefix "C".

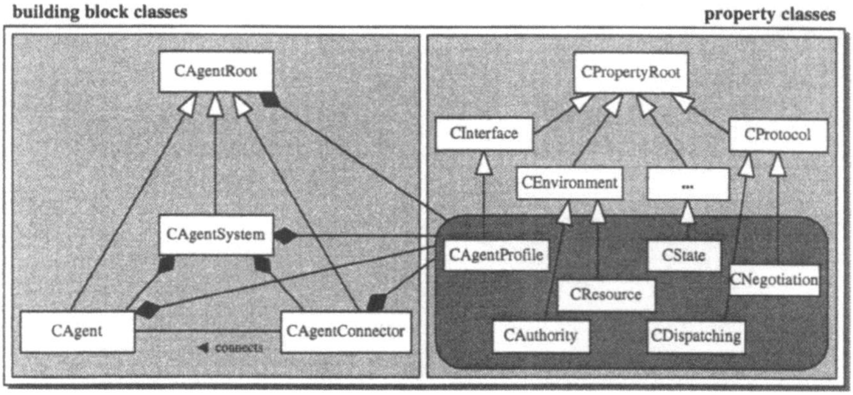

Figure 1: Basis architecture type for agent systems in UML notation

The property classes have been classified (Horn, Kupries and Reinke 1997) on the basis of the available scientific literature, existing standards, implementation platforms for agent systems and practical applications. These analyses allow implementation and object-oriented modelling of properties for agent system building blocks. The result is a hierarchy of property classes for the building blocks of the architecture type for agent applications. It is depicted on the right-hand side of *figure 1*. The separation of the property classes from the building block classes allows their separate prefabrication and reuse.

2.3. ARCHITECTURE TYPES FOR APPLICATION DOMAINS

Further analysis of multi-agent systems for typical applications (like information retrieval and electronic commerce) allows the abstraction of application-specific building blocks and their properties. It also allows the development of architecture frameworks for these classes of application systems. The definition of architecture frameworks is very well supported by object-oriented modelling with its possibilities for abstraction, construction of subsystems and so on.

The described application-specific building blocks are generally composite. They can be derived from or composed of elementary building blocks that are independent of applications. It is possible to arrange the building blocks and their properties hierarchically in libraries. Furthermore a similar hierarchy can be derived according to the architecture types. Within the hierarchy, an increasingly close relation to applications can be observed (*figure* 2). This figure shows some of the classes of the Basic Agent System Architecture Type, the Abstract Application Architecture Type and an Architecture Type for the Market Place application domain. This serves as a preview to the sample architectures discussed in section 3. The inheritance relations between the classes are sketched. For reasons of clarity we avoid depicting other relations in *figure 2*.

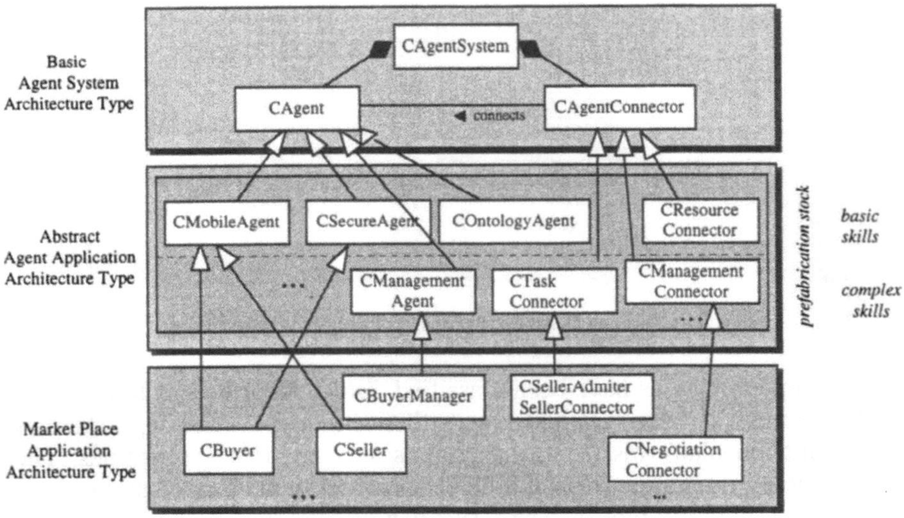

Figure 2: Hierarchy of architecture types

3. Agent applications and sample architectures

On the basis of the derived classes (*figure 2*), specimens of architecture frameworks can be derived as a framework for agent application systems. Using the common software technology terminology, they are called sample architectures. Sample architectures have to be interpreted and specialised for a concrete application. They form an important basis for the creation of development methods in software agent systems.

3.1. COMMON AND SPECIFIC SAMPLE ARCHITECTURES

The common sample architecture for agent applications depicted in *figure 3* has emerged as a result of the analysis of various typical applications from the fields of electronic commerce and information retrieval, such as Carminelli-Gregori, Cobalti and Vercelli (1999), Chavez and Maes (1996) and Rodriguez, Noriega,

Sierra and Padget (1997).

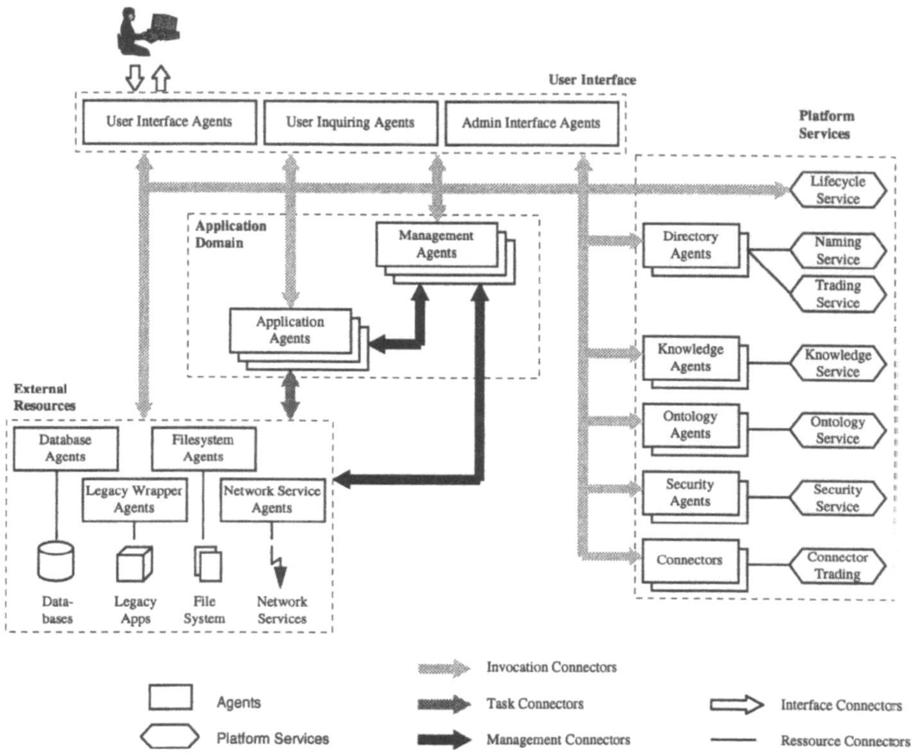

Figure 3: Common sample architecture for agent applications

The building blocks are grouped into the subsystems:

- user interface
- application domain
- platform-specific services
- external resources

The **User Interface** subsystem includes agents that take over communication with the user via interface connectors. Three kinds of software agents can be differentiated in this subsystem. User interface agents take over communication with the user to specify the domain-specific tasks of the agent system. User Inquiry Agents take over communication with the user for system and user controlled queries while an agent system is dealing with a task. Administration

Interface Agents realise user communication with the system services.

The **Platform Services** subsystem comprises agents that provide access to agent specific services. There is special emphasis on the Ontology Service that plays a crucial role in open agent systems. This subsystem also includes the Knowledge Service containing knowledge about the application domain which is based on the ontology. Unlike distributed problem solving in artificial intelligence, the starting points are not knowledge bases (central and local) describing the application domain completely (no automatic collective problem solving). The knowledge base of an application domain develops with the number of problems solved. Incompleteness is avoided by the fact that the user is prompted (via a User Inquiry Agent) if knowledge and ontology agents cannot solve a problem with the appropriate services.

The platform services also include connector trading. Having described a domain specific sample architecture, some features of the functioning of connectors are explained in section 3.2.

The **External Resources** subsystem contains agents regulating the access to data bases, file systems, external applications and network services via resource connectors.

The **Application Domain** subsystem contains the agents and connectors for the actual application domain and has to be further specialised and derived. To my mind, the agents and connectors of the other subsystems are less application specific.

The common sample architecture is the starting point for a specialisation towards sample architectures for specific classes of agent applications. By refining the subsystem **Application Domain**, for instance, it is possible to derive a sample architecture for an electronic market place in which agents offer or auction products for their authorities (*figure 4*).

In this figure the connectors are assigned numbers indicating the connection sequence during the work of the market place. This sample architecture can be used for all agent systems of this kind regardless of the products that are traded. The sample architecture reflects a particular organisational structure (offices for stationary agents). On the basis of the architecture type depicted in *figure 2*, sample architectures for other organisational structures can also be derived. In this sense sample architectures are the frameworks for agent application systems. The same is true for the common sample architecture.

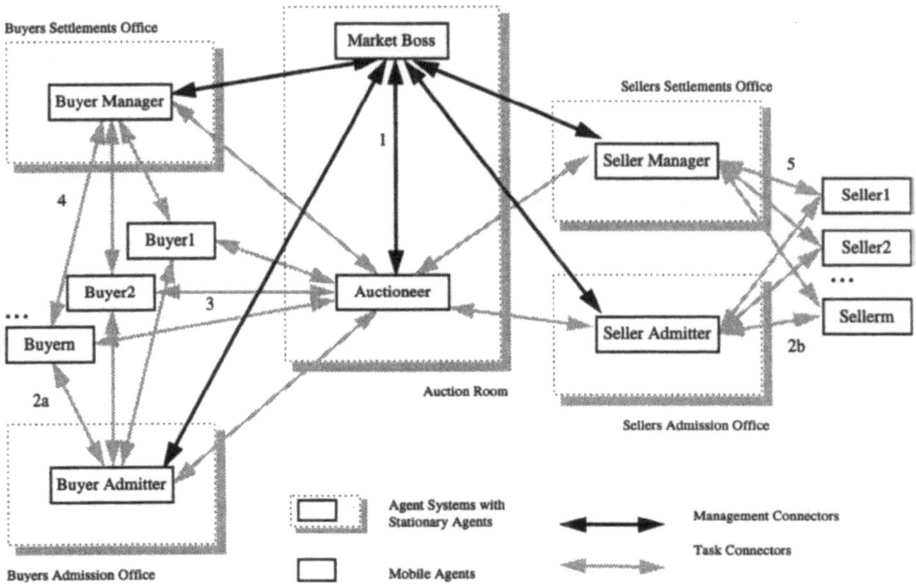

Figure 4: Sample architecture for the application domain Electronic Market Place

Other application specific sample architectures that have been defined by the abstraction of corresponding applications are information retrieval on the net and information logistics in medical applications and for a travel agency.

4. Building blocks in sample architectures

In the sample architectures, the building blocks are instances of agent classes and connector classes. The objects of prefabrication are agent classes, connector classes and sample architectures. The agent classes and connector classes of the sample architectures have different degrees of application dependency. The agent classes and connector classes of the common sample architecture are less application specific than the sample architectures for application domains are.

To explain the necessity of explicit modelling and the functioning of connectors, we shall have a look at the connector which takes over communication between seller and seller admitter (*figure 4*). The functioning of this connector is explained in *figure 5*.

Connectors are explicit building blocks of software architectures. During run time an agent asks (via the Connector Trading) for a connector with a particular profile for communication with another agent.

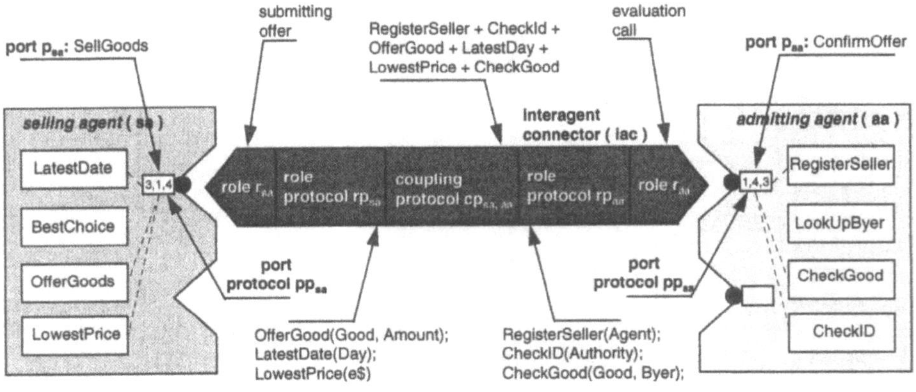

Figure 5: Example of a connector according to Kupries (1999)

An instance of the appropriate connector class is generated. Three connector life cycle phases can be differentiated: connection establishment, communication and disconnection.

During connection establishment, the matching of the connector roles with the agent ports is checked. The agent ports contain agent interface functions and, where appropriate, sequence rules. An agent can have several ports. A connector can connect two agents with each other if its roles correspond with the agent ports.

Having been coupled, agents can communicate with each other via a connector according to a protocol which is controlled by the connector (cf. *figure 5*, coupling protocol). Having finished communication, the connection is terminated.

This example shows that the connectors can be application specific. That is why a common connector as part of the agent platform does not meet the requirements. Connectors can have various protocols. There are also connectors that deal with the treatment of results or which can arrange negotiations between two agents. Thus connectors, just like agents, are building blocks of the agent system and objects of prefabrication.

There are different degrees of abstraction for the description and prefabrication of the building blocks (agents, connectors and sample architectures). Such forms include:

- prefabricated source texts
- pattern-like descriptions
- model types and corresponding tools

Prefabricated source texts can be developed for agents and connectors. They can be managed in libraries. Generic concepts and frameworks can also be used.

The pattern-like description of building blocks can be realised using a scheme developed with reference to description schemes for design patterns (Gamma, Helm, Johnson and Vlissides 1994, Buschmann, Meunier, Rohnert, Sommerlad and Stal 1996). The scheme specifies the properties of the building blocks, their relations to other building blocks and the environment, their behaviour, the informal description of their semantics and the integration of building block and application specific security properties. Furthermore, schemes describe the possibilities and consequences of building block use in applications. Examples of such descriptions are to be found in Reinke and Horn (1999). Patterns are abstracted knowledge. The prefabrication of building blocks for an agent system is done on the basis of patterns.

A model type describes structural properties of a set of models and operations for the construction, alteration and assessment of models as specimens of the model type. A corresponding tool supports the construction, alteration and application of specimens of the model type. A clear example of this is the ontology agent and the Ontology Service. Following the FIPA definition (FIPA 1998), the ontology of an application domain is constructed as a term model, i.e. it contains the terms of the application domain and their relations. The model type (often referred to as the metamodel) of the term model (described by means of expressions in UML) is shown in *figure 6*.

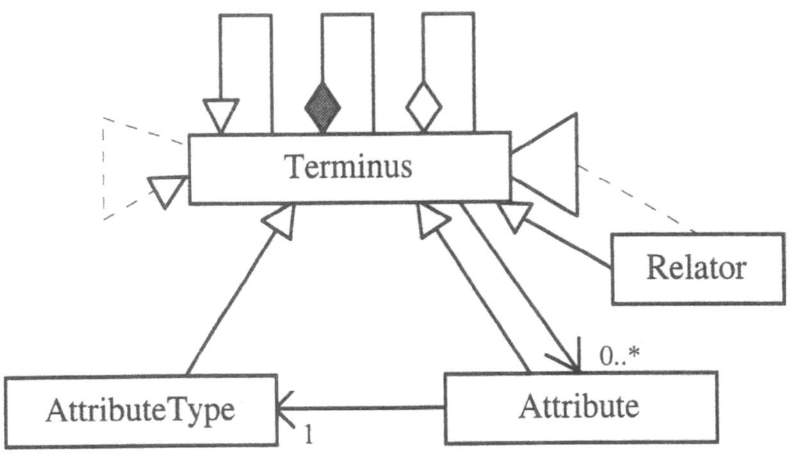

Figure 6: Model type of the term model

The model type illustrates that the ontology consists of attributed terms that can be related by abstraction, composition, semantics, use and instancing. The ontology for an application domain is a specimen of this model type.

4.1. SAMPLE ARCHITECTURES AND DEVELOPMENT METHODS

Sample architectures offer a number of benefits during the development of methods for the construction of multi-agent systems. These can be summarised as follows:

- Sample architectures are abstracted knowledge about the architecture for classes of application systems
- The subsystem structure of the sample architecture reduces complexity and allows specialisation of parts to emerge
- Sample architectures allow the reuse of building blocks
- Agent application systems can be developed "into" sample architectures. For the development of the individual parts, special models and means of expression which are characteristic of subsystems can be used
- The development of products into sample architectures is a traditional scientific-engineering approach

5. Aspects of distributed artificial intelligence and of sociology in multi-agent systems

5.1. SOME REMARKS ON THE RELATIONS TO DAI

A close affinity to distributed artificial intelligence is especially obvious in the Ontology and Knowledge Services. To explain this, we refer to work dealing with the ontology of the product "Car" for the market place application domain (Alberts, Jonker, Karami and Treur 1999). A sample architecture for the application domain is shown in *figure 4*. The ontology for this application domain is subdivided into several parts according to the available roles (action scope of the agents):

- Basic Product Ontology
- Provider Perspective Ontology
- Customer Perspective Ontology.

A part of the basic product ontology is shown in *figure 7*. One can see that this model is a specimen of the model type term model.

Unlike Alberts, Jonker, Karami and Treur (1999), we do not look at rules as part of the ontology but we correlate them with the knowledge base.

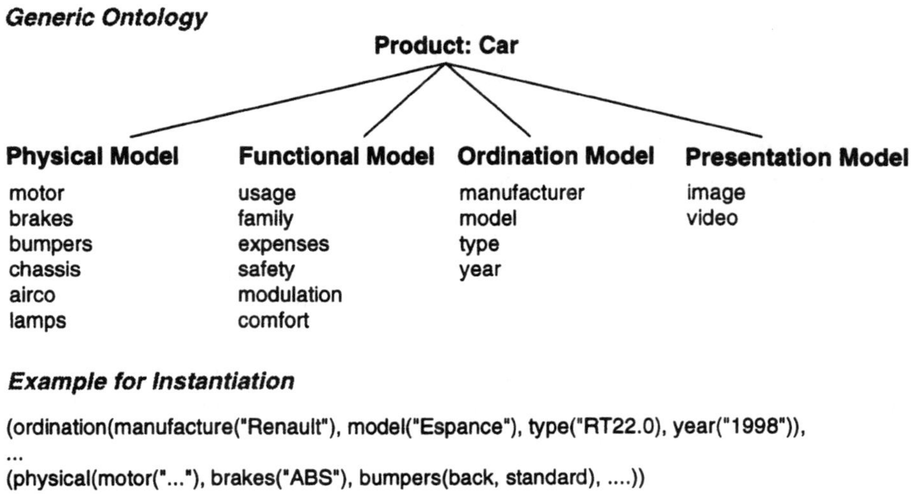

Figure 7: Example of the ontology with reference to Alberts *et al.* 1999

The knowledge base contains rules that refer to the properties of the product "Car", e.g. *figure 8* shows a rule for the security assessment of a car.

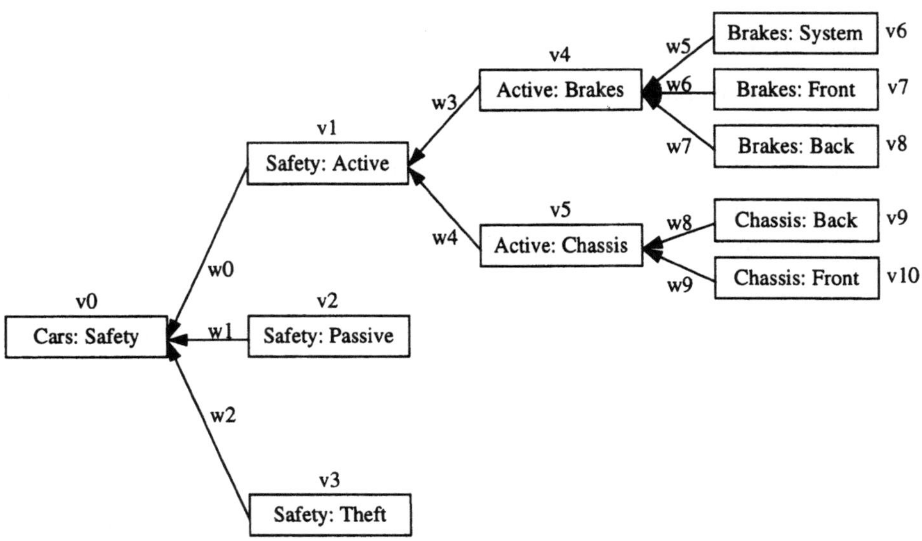

Figure 8: Rule for car security assessment according to Alberts *et al.* (1999)

The product properties and their relations are assessed with weighting factors in this rule. Before the application of the rule, the weighting factors have to be bound. The knowledge base contains the rules necessary for the work of the agent system. If a new rule is needed, it has to be entered via the Platform Service into the knowledge base during system run time. The maintenance of consistency and integrity in the knowledge base are difficult in this respect. Suppose the agent system is to assess an aspect and does not find any rules in the knowledge base. It must establish (via the corresponding User Inquiry Agents) a connection with the person who does the assessment in the dialogue or who has to generate the appropriate rule and enter it into the knowledge base. From the point of view of software engineering, these are systems that do not have a "world model" and are thus dependent on communication with human beings. Interdisciplinary work between DAI specialists and those from software engineering is necessary to establish the ontology and knowledge base.

At present, one chiefly encounters agent application systems with an ontology and knowledge base that cannot be varied. This is, of course, easier and faster to implement but does not support openness, expandability and polyglotism in agent systems.

Another problem with the development and use of agent systems is the neglect of important sociological aspects of such systems. These are becoming more and more important from the perspective of software engineering because multi-agent systems have to be regarded as part of the more general class of man-machine systems. Here, too, interdisciplinary work between sociologists and computer scientists is urgently required. The foci for cooperation stem particularly from the following requirements:

- Sample architectures for agent systems have to be close to the real human systems in cognitive terms
- The ontology should be selected according to sociological criteria
- Agents for the User Interface subsystem need to be implemented in a human way.
- Interface connectors have to be adaptable to human beings
- Agent systems have to be transparent and trustworthy for human beings
- Unnecessary technical details of the system have to be hidden
- The knowledge should be distributed efficiently between human beings and agent systems
- One should aim at "learning" agent systems in which human beings progressively accumulate knowledge in the agent system
- The roles in agent systems should be adapted to the real world
- Social contacts and groups have to be emulated in agent systems

6. Advantages and disadvantages of agent systems and outlook

Agent systems (seen from the point of view of software engineering) have reached the transition stage between experimental and industrial applications. However, the absence of standardised platforms for the development of agent systems is an ongoing problem. The platforms for mobile agents available in industry today (e.g. IBM Aglets and Object Space Voyager) are still being developed. There are many common and special platforms that have been developed in universities and research institutes (e.g. MOLE). As a rule, the systems are influenced by views of communication technology. They do not support a development technology. However, a platform (*A*gent *A*rchitecture Type-Based *D*evelopment *E*nvironment - AADE), supporting architecture types is currently being developed at Potsdam University. In practice, "informal" Java-based implementations are prevalent in the field.

With all the technological euphoria, one should have a clear idea about the advantages and disadvantages of agent systems.

Advantages of agent systems are mobility, autonomy (the decentralisation aspect), flexible use and the individual service of agents, net-wide openness and the dynamics that goes with it and net load reduction when using the agent system properly. In addition, there are possibilities for varied interaction and human-like communication in agent systems, ability to negotiate, domain specific ontology, openness at run time and scalability of the agent system.

Disadvantages of agent systems are higher security requirements, high development costs, the possibility of redundant information in the system, a low degree of availability and experience (at least until now), missing development technologies and tools, and insufficient experience of heterogeneous environments.

The development of agent application systems is demanding. It requires a knowledge of the Java programming language, of class libraries and agent libraries, of network communication, security mechanisms, man-machine communication, ontology definition and so on. When designing agent application systems, one has to take many different aspects into consideration.

Without re-using knowledge (in the form of sample architectures and building blocks) and without interdisciplinary cooperation neither efficient development nor high quality agent application systems will be possible. We would like to emphasise at this stage that agent technology is really efficient where programs solving a special task are small and the remote data bases required are large.

All in all, we can expect that big application systems will be heterogeneous, using the technological approaches of both local and distributed programming. Agent technology will play an important role in this process but will have to integrate itself into the set of other technologies.

References

Alberts, M., Jonker, C., Karami, M., and Treur, J. (1999). *An Electronic Market Place: Generic Agent Models, Ontology and Knowledge; Autonomous Agent.* Seattle WA . 19-21.

Buschmann, F., Meunier, R., Rohnert, H., Sommerlad, P., and Stal, M. (1996). *Pattern-oriented Software Architecture: A System of Patterns.* John Wiley, Chichester.

Carminelli-Gregori, A.-M., Cobalti, R., and Vercelli, G. (1999). *Intelligent Web Agents for Information Retrieval and Classification.* Fourth International Conference on the Practical Application of Intelligent Agents and Multi-Agent Technology. London.

Chavez, A., and Maes, P. (1996). *Kasbah: An Agent Marketplace for Buying and Selling Goods.* First International Conference on the Practical Application of Intelligent Agents and Multi-Agent Technology. London.

FIPA 98. Foundation of Intelligent Physical Agents, http://www.fipa.org, 1998.

Gamma, E., Helm, R., Johnson, R., and Vlissides, J. (1995). *Design Patterns: Elements of Reusable Object-Oriented Software.* Addison-Wesley, Reading, MA.

Horn, E., Kupries, M., and Reinke, T. (1997). *Properties and Models of Software Agents and Agent Systems*; Universität Potsdam Preprint 006/1997; ISSN 0946-7580.

Horn, E., Kupries, M., and Reinke, T. (1998). *Architekturtypen für Agentenanwendungssysteme*; Universität Potsdam Preprint 001/1998; ISSN 0946-7580.

Horn, E., Kupries, M., and Reinke, T. (1998). *Object-oriented software architecture types for the substantiation, development, prefabrication of agent application systems.* Eleventh International Conference on Software Engineering and its Applications. Paris.

Horn, E., Kupries, M., and Reinke, T. (1999). *Properties and models of software agents and prefabrication for agent application systems.* 32nd Hawaii International Conference on System Sciences, Hawaii, 1999.

Horn, E., and Schubert, W. (1993). *Objektorientierte Software-Konstruktion.* Grundlagen, Methoden, Modelle, Beispiele. Hanser Verlag, München.

IBM Aglets: http://java.trl.ibm.com/aglets

Iglesias, C., Garijo, M., Gonzales, J., and Velasco, J. (1998). *Analysis and Design of Multiagent Systems using MAS-CommonKADS.* Intelligent Agents IV: Agent Theories. Architectures and Languages. Springer-Verlag, Berlin.

Kinny, D., and Georgeff, M. (1997). *Modelling and Design of Multiagent Systems.* Intelligent Agents III: Agent Theories, Architectures and Languages. Springer-Verlag, Berlin.

Kupries, M. (1999). *Connector-aided Coordination in Agent Systems.* Coordination in Parallel and Distributed Applications and Activities. Las Vegas.

Mobile Agent Systems Interoperability Specification. Object Management Group, OMG Document orbos/98-03-09, 1998.

MOLE: http://www.informatik.uni-stuttgart.de/ipvr/vs/projecte/mole.html

Object Space Voyager: http://www.objectspace.com/Voyager/voyager.html

Reference Model for Frameworks of Software Engineering Environments, NIST, Gaithersburg, MD, 1993.

Reinke, T., and Horn, E. (1999). *Musterarchitekturen und Bausteine für Multiagentensysteme.* Universität Potsdam, Institut für Informatik, Technischer Bericht, 1999.

Rodriguez, J., Noriega, P., Sierra, C., and Padget, J. (1997). *FM96.5 A Java-based Electronic Auction House.* Second International Conference on the Practical Application of Intelligent Agents and Multi-Agent Technology. London

Rumbaugh, J., Jacobson, I., and Booch, G. (1999). *The Unified Modelling Language Reference Manual.* Addison-Wesley, Reading, MA.

System Application Architecture: An Overview, GC26-4341, IBM Corperation, 1989.
Shaw, M., and Garlan, D. (1996). *Software Architecture. Perspectives on an Emerging Discipline.* Prentice Hall, New York.
Wooldridge, M.J., Jennings, N.R., and Kinny, D. (1999). *A Methodology for Agent-Oriented Analysis and Design.* Autonomous Agents. Seattle.

SOCIAL NORMS FOR CO-OPERATIVE AGENTS

NICOLE J. SAAM
Institute of Sociology
Ludwig-Maximilians-Universität München
Munich, Germany

Abstract

The importance of more precise (non-metaphorical) social notions (such as norms) in Distributed Artificial Intelligence has long been emphasised. Co-ordination and common problem-solving among agents could require equivalents to social norms as they can be observed in human societies. However, these phenomena should not be intentional or planned but "emergent". This paper investigates social norms for co-operative agents from a sociological point of view. It outlines how sociological theory defines social norms, explains their emergence in human societies and explores the extent to which empirical social research has been successful in verifying their efficacy. It then suggests concepts of norms and norm processing that could either be used for planned implementation or for identifying the equivalents of social norms among co-operative agents. Finally, it assesses how DAI and sociology could profit from each other's research in norms.

1. Introduction

The importance of more precise (non-metaphorical) social notions (such as norms) in Distributed Artificial Intelligence has long been emphasised. (See Castelfranchi and Werner 1994, xii, and Müller, Malsch and Schulz-Schaeffer 1998, 1.1.) In sociology, social norms are seen as central mechanisms of co-ordination and co-operation among human beings. Social norms, as solutions to recurring problems of social interaction, are seen both as results and preconditions for social life, unintended outcomes and human devised constraints. Coordination and common problem-solving among co-operative software agents could require equivalents to social norms as they are observed in human societies. However, these phenomena should not be intentional or planned but "emergent" (Castelfranchi and Werner 1994, ix). Sociological theories define social norms and explain their emergence in human societies while empirical social research investigates the efficacy of social norms. It should be possible, then, for sociology and DAI to co-operate in research on social norms. Sociology starts with a serious evaluation of sociological conceptions and theories for computer systems (Malsch 1997) while DAI

N.J. Saam and B. Schmidt (eds.)
Cooperative Agents. Applications in the Social Sciences, 39-56

applies more or less sociologically founded conceptions and reveals their limitations or unobserved implications (Conte and Castelfranchi 1995b).

This paper investigates social norms for co-operative agents from a sociological point of view. It outlines how sociological theories define social norms and explain their emergence in human societies (section 2.1). It also considers the extent to which empirical social research has been successful in verifying the efficacy of social norms (section 2.2). We describe the conclusions that can be derived from research on norms among co-operative agents in DAI (section 2.3). Then, we suggest concepts of norms (section 3.1) and norm processing (section 3.2) that could either be used for planned implementation or for identifying the equivalents of social norms in populations of co-operative software agents. Finally, we assess how DAI and sociology could profit from each other's research in norms (section 4).

2. Social norms in sociology

The study of social norms is a classical research topic in sociology. The sociological study of norms aims at the appropriate definition of norms, the explanation of ways in which norms affect social behaviour, and processes by which norms emerge, become established (and internalised) and change.

2.1. SOCIAL NORMS AND SOCIOLOGICAL THEORY

2.1.1. Conceptualisation of norms

The sociological study of norms has developed four different conceptualisations of norms (Nonner-Winkler 1984) including a dismissal of the concept of norms itself:

(1) The *statistical conceptualisation of norms* originates in behaviourism. A behavioural pattern emerges as a norm if the majority of actors in fact behave according to this pattern. Thus, norms are objectively observable and measurable. From this perspective any regular behaviour becomes a norm. One cannot distinguish between behaviour that is regular and rule obeying, conditioned and rule obeying or sanction avoiding and rule obeying. Theories of social norm formation (proposed by Sherif 1967 and Geiger 1962) are examples of this approach.

(2) Theories of action from Durkheim, Weber and Parsons represent *the "sociological" conceptualisation of norms*. According to Durkheim, norms are social facts which can be identified through the mere existence of certain sanctions (1965, 112). The sanctions do not constitute the norm, they are symbols of it "and as this symbol has the great advantage of being objective, accessible to observation and even to measurement, it is a good method to prefer it to the thing it represents" (1966, 426). Durkheim does not deny the

reality of a subjective perspective on norms, but he promotes the methodological principle that scientists should choose the perspective of an objective observer. It was Weber (1960) who was responsible for making the distinction between objectively observed behaviour and subjectively intended action explicit. There are behavioural regularities that are merely based on practice (customs or traditions), on similar interest (market behaviour) and on notions of legitimate order. It is constitutive of normative action that the actor perceives behavioural rules as obliging or representative (ibid. 26). In the theory of Parsons (1951), norms represent institutionalised role expectations. The objective character of norms is guaranteed by the integration of role expectations into the cultural system and its subjective character by internalisation.

(3) Ethnomethodologists deny the overall relevance of the sociological conceptualisation of norms. According to Garfinkel (1967), there is no such thing as a general norm independent of concrete contextual conditions. There are only situation specific expectations of concrete interaction partners that are actively negotiated by both sides. There is then a question about how individuals constitute a common world. Ethnomethodologists have discovered several basic rules which have a pseudo normative character. They are seen as obliging and deviations are sanctioned. But they are not equivalent to sociological norms. They are far more deeply rooted in sociality and are not amenable to reflection. Deviation from these base rules is judged according to clinical and not moral categories. Furthermore, "a society's members know the moral order as perceivedly normal courses of action" (Garfinkel 1967, 35). In practice, actors resort to the statistical conceptualisation of norms in concrete interactions.

(4) Finally, developmental psychologists have put forward an *ethical conceptualisation of norms*. Piaget (1983) and Kohlberg (1971) attempt to reconstruct the development of human moral consciousness. At any time, a subject is actively structuring and reflecting his or her social environment. Three main stages can be distinguished in the subject's moral development. The first stage is pre-conventional morality: a child complies with norms in order to avoid punishment. This corresponds to behaviourism. The second stage is conventional morality. A subject complies with norms because they have internalised the norms of the family, peer group or society. This corresponds to the Parsonian view. The third stage is post-conventional morality. A subject complies with norms because of insight into abstract principles that allow for the foundation and justification of norms. This corresponds to the ethical conceptualisation of norms. At this stage, the subject is able to distinguish between statistical norms, "sociological" norms and ethical norms.

2.1.2. The emergence of norms

Rational choice theory turns to the question of whether there is a basic contrast between rationality and social norms. (See Coleman 1986, 1986/87, 1990a, 1990b and Elster 1989a, 1989b, 1991.) Game theory has developed three different conditions for the emergence of norms among rational actors (Ullman-Margalit 1977):

(1) Norms as solutions to problems of co-ordination: Here, norms are seen essentially as conventions, behavioural conformities that do not presuppose explicit agreements among agents. They emerge from the individual interests of the agents (Schelling 1960, Lewis 1969) but do not presuppose explicit agreements among them. Lewis' example is that of two persons unexpectedly cut off during a telephone conversation. Each has to choose whether to call back or wait. Norms of coordination emerge gradually from interactional practice. There is no conflict of utility because both participants want to resume their conversation.

(2) Norms as solutions to conflicts over utility: In the case of social dilemmas, the conflict structure calls for norms whenever the agents' choices tend to produce a state of affairs that is both individually and socially undesirable.

(3) Norms as solutions to problems of inequality: In this case norms aim at maintaining a state of affairs in which one party is favoured or privileged at the expense of another.

2.1.3. The functions of norms

Functional analysis seeks the objective effect that one element of a social system produces for the system as a whole. There is not necessarily common knowledge about the objective effect of a social element. Functional analysis in general does not concentrate on where the causal relation lies. Cause and effect relations are a special case of functional relations. Different causes are functionally equivalent if they produce the same effect.

Right from the very beginning, the functions of norms in society have been a central field of research in sociology. There are two positions in this debate. Firstly, norms have been analysed as solutions to social problems. Marx (1936) considers norms to be dependent on the economic foundations of society. Accordingly, the functions of norms follow the interests of the ruling class. According to Durkheim "society cannot exist if its parts are not solidary" and "morality consists in being solidary" (Durkheim 1966, 399). In the Parsonian L-I-G-A or A-G-I-L scheme, norms fulfil the integration (I) function of social systems. Solidarity is produced primarily at the level of values which are shared by more or less everyone in society. Children are socialised by their parents so that the basic values and norms of society are "internalised" as parts of their personalities. As a result, individuals need and want to do what society demands of them. Alternatively, norms have been analysed as generating social problems themselves. Dahrendorf states that wherever behaviour is regulated by established norms, and wherever these norms are backed by sanctions, a rank order of social status is bound to emerge (Dahrendorf 1962, 102).

Computer simulations of norms in artificial societies have explored three different functions of norms. Firstly, to permit or improve co-ordination among the agents (Shoham and Tenneholtz 1992a, 1992b). Secondly, to control aggression and thirdly to reduce inequality (Conte and Castelfranchi 1995a, Castelfranchi, Conte and Paolucci 1998, Walker and Wooldridge 1995). It has been demonstrated that societal evolution from egalitarian to non-egalitarian societies may lead to functional change in norms. Whereas property norms reduce inequality in egalitarian societies, they increase inequality in non-egalitarian societies (Saam and Harrer 1999).

Lindenberg (1985) has identified two types of a normative "homo sociologicus" which have been investigated by the majority of sociologists: the *socialised, role-playing, sanctioned man* (SRSM) and the *opinionated, sensitive, acting man* (OSAM). Both assume that men comply with norms they have internalised. While the first type refers to sociological theorists like Durkheim, Weber and Parsons who represent the "sociological" conceptualisation of norms, the latter conception refers to the majority of empirical social researchers.

2.2. SOCIAL NORMS AND EMPIRICAL SOCIAL RESEARCH

Typically, empirical social research asks for the attitudes and modes of behaviour that individuals show in certain situations. The attitudes comprise norms and values that are held by the individuals. Most often, the attitudes are assumed to be exogenous variables, and they are used to explain endogenous variables like the behaviour of the individuals. Numerous statistical procedures are available to test the interrelation between exogenous and endogenous variables. It is the general intention to explain the behaviour of the individuals with the help of their attitudes. This is the second type of a normative "homo sociologicus", the *opinionated, sensitive, acting man* (Lindenberg 1985). However, the explained variance of these causal models from empirical social research is generally quite low (Esser 1993, 234). It varies between 20 and 50%. As Esser points out, the relations between attitudes and behaviour are always very weak and indirect. Even if one variable seems to affect another (as indicated by a high beta coefficient) there is still a long way to go before "understanding" the mechanisms that produced these behaviours. Behaviour that deviates from norms and attitudes seems to be the rule rather than the exception. Compliance with norms (and not deviance) seems to be the social fact that requires explanation in complex societies (ibid. 234). Furthermore, under conditions of uncertainty, conflict, ineffectiveness of socio-demographic variables and the absence of all-embracing orientations, the normative "homo sociologicus" provides little explanatory power.

2.3. CONCLUSIONS

In sum, there are quite different conceptions of norms in sociology. The functions of social norms are relative. They depend on time and space, like so many other social facts. The variance that causal models from empirical social research explain is unproductive. The relations between attitudes and behaviour are always very weak and indirect.

Obviously, the idea of simply transferring sociological norm concepts to DAI is less promising than it seemed before. Castelfranchi and Werner's claim that social norms should emerge among co-operative software agents, but not be intentionally implemented or planned, is supported by these findings (Castelfranchi and Werner 1994, ix). Social science could offer DAI "many important, puzzling, and well studied but unresolved problems of social interaction ... and some relevant formal approaches and theories about human action..." (ibid. xiii). These theories (including rational decision theory, economic theory and game theory) should be used and absorbed critically. It should not be forgotten that they are under debate (from both a philosophical and an empirical point of view) in their original disciplines (ibid. xiii).

In the following discussion, we propose concepts of norms and norm processing suitable for application to co-operative software agents. They could (but need not be) used for planned implementation. It might be more promising to use them to identify equivalents of social norms that have emerged unintentionally among co-operative software agents.

3. Social norms for co-operative agents

In computer science, the terms "agent" or "intelligent agent" are technical but do not have a generally agreed usage. An agent denotes hardware or (more usually) software based computer systems. According to Wooldridge and Jennings (1995, 2) a weak notion of agency may be distinguished from stronger notions. The properties of a weak agent are autonomy, social ability, reactivity and pro-activeness. The stronger (mentalistic) notion includes properties, such as knowledge, belief, intention and obligation (Shoham 1993). Bates (1994) even considers emotional agents. Further attributes are flexibility, planning, prediction, selectivity, robustness and learning ability (Moffat and Frijda 1995, 246) as well as mobility, veracity, benevolence and rationality (Wooldridge and Jennings 1995, 2).

Sociologists may contribute to understanding of several of these features. Carley and Newell (1994) describe social agents along two dimensions: processing capabilities and differentiated knowledge. We concentrate on the interactivity of agents (that is on their social ability). If an agent is a member of a society it must be able to interact with other agents. Interactivity implies communication, coordination and co-operation between agents (Brenner, Zarnekow and Wittig 1998, 25f). Here, again, we concentrate on coordination and co-operation. We are only interested in coordination and co-operation among intelligent agents. To indicate this, we call our agents co-operative agents. Co-operative agents co-operate in problem solving. In distributed systems, each agent solves one part of a problem. After each agent has solved its part of the problem the partial solutions have to be integrated to produce an overall solution.

As in human societies, norm processing could be one mechanism that co-operative agents use to manage coordination and co-operation. For example, software agents that assist their human principals in buying or selling products on a virtual market will have to co-operate with each other. These agents are expected to become very important in electronic commerce (Brenner, Zarnekow and Wittig 1998, 308ff). Sooner or later, such agents will not only be capable of simple bargaining. Virtual markets are being given economic constitutions. Agents, as well as their human principals, will expect that other agents comply with (legal) norms that are part of these economic constitutions. Nevertheless, some agents will deviate. To maintain economic order, they will have to be sanctioned. A legal system will be needed to guarantee long term economic constitutions. We also expect the evolution of social norms. As Ullman-Margalit (1977) shows social norms emerged in human societies as solutions to problems of co-ordination, to conflicts over utility and to problems of inequality. Social norms will probably also be efficient in solving these problems among agents. From the sociological point of view, legal norms are a special case of social norms. Processing norms, therefore, seems to be a fundamental requirement of co-operative software agents.

The advantages of norm governed systems are twofold. They avoid useless, stupid and self-destructive behaviour which is favoured by the rigid execution of routines. They also prevent the spreading of errors and deviations produced by pure imitation (Conte and Castelfranchi 1995b, 91). Therefore, computer scientists ought to be interested in agents with a capacity for processing norms. However, computer scientists will only accept norm processing agents if this approach increases program productivity: algorithmic efficiency (computing time and memory requirements), elegance in system architecture and software maintainability.

In the following sections, we suggest concepts of norms and norm processing appropriate for co-operative software agents. They could (but need not be) used for planned implementation. It might be more promising to use them to identify the equivalents of social norms among co-operative software agents which have emerged unintentionally. To develop (or identify) co-operative agents that have norms processing capacities, one has to advance step by step:

1. Develop norm concepts for software systems
2. Develop conceptions of norm processing
3. Determine the contents of prescribed norms
4. Develop conceptions that allow norms to change their contents

In the following discussion, we concentrate on steps one and two.

3.1. NORM CONCEPTS

Following Ullman-Margalit (1977), Conte and Castelfranchi (1995b, 77) refer to the work of the moral philosopher Hart (1961). He defines norms as "a prescribed guide for conduct or action which is generally complied with by the members of a society". Three features are characteristic of norms: their prescriptive force, their expected necessity for social life and the inevitable clashes between norms and the interests and desires of the agents who owe the duty. The second feature is criticised by Conte and Castelfranchi as vague. Hart himself states "The rules supported by this serious pressure are thought important because they are believed to be necessary to the maintenance of social life or some highly prized feature of it" (ibid. 85). Conte and Castelfranchi propose a model in which norms are seen as both macro- and micro- objects, social facts and mental constructs. Two concepts of normative behaviour are introduced:

1. Norm-abiding behaviour, which appears to correspond to some norms
2. Norm-governed ("fully normative", Conte and Castelfranchi 1995b, 90) behaviour, which results from cognitive processing of norms and implies a mental representation of norms

Conte and Castelfranchi put forward the thesis that norm-governed systems are more useful than norm-abiding systems. The extent to which norm-governed or norm-abiding systems perform better needs to be tested. They state that "all of these hypotheses should be tested computationally. Unfortunately, we are far from being able to execute such a test" (ibid. 92). They also discuss the relative advantages of three different architectures of norm-abiding behaviour. Norms as built-in constraints, goals or obligations differ in their reliability, learning capabilities, novelty, faculty for change (and repair) and effectiveness in social control. Each conception makes the representation of norms increasingly demanding, leading ultimately to norm-governed behaviour (see *table 1*). Most important, a built-in obligation differs from a constraint, since it does not automatically reduce the set of actions available to the agents. It is also different from a goal, since it is not immediately transformed into a goal (ibid. 92).

Norm concept	built-in constraint	built-in end (goal)	built-in obligation
definition	norms are seen as constraints on the agents' action repertoires (a behavioural notion); no truly normative choice	cognitive agents are allowed to choose among competing goals (instead of simply applying procedures and routines) but they treat norms like any of their other goals	explicit and specific mental object; differs from a constraint since it does not automatically reduce the set of actions available to the agents; and from a goal since it is not immediately transformed into a goal
reliability	high reliability: the constraints will always be executed	low reliability: goals may be abandoned when they clash with competing, more urgent needs	average reliability: agents may give up normative goals, but this will cost them more than abandoning ordinary goals
learning	no learning: new constraints are implemented when the system is off-line	average learning: built-in ends should be added when the system is off-line	normative goals are autonomously produced on the grounds of normative beliefs and in principle beliefs may be acquired when the system is on-line
novelty	no novelty: constraints reduce available actions; only proscriptions, or prohibitions ("Don't ...") can be represented and not prescriptions ("Do...")	built-in ends may correspond to prescriptions and not only to prohibitions; but since learning is not granted, there ought to be a one-to-one correspondence between social norms and internal goals with a consequent high computational complexity	normative beliefs may be formulated as prescriptions and not as mere prohibitions
repair	no repair: agents are not endowed with the capacity for modification of unsuccessful constraints	goal-oriented systems ought to be able to try out several solutions and choose those that achieve their goals to the highest degree, before giving up their goals	as with any other kind of goal, plans for normative goals are subject to change and repair
social control	no need	no social control: agents may have preferences corresponding to norms or not; they may cling to these or abandon them as a function of circumstances	since normative goals are a special type of goals, and normative actions have specific costs, agents can be shown to be interested in some monitoring of norms

Table 1: Alternative representations of norms inside the agent from Conte and Castelfranchi (1995b, 92)

Until now, norm have usually been operationalised in the Distributed Artificial Intelligence literature as behavioural constraints on agents. At present, the computational development of concepts (Shoham and Tenneholtz 1992a, 1992b, Conte and Castelfranchi 1995a, 1995b, Walker and Wooldridge 1995, Castelfranchi, Conte and Paolucci 1998) is moving from the statistical to the "sociological" conceptualisation of norms. (See especially Conte and Castelfranchi 1995b, Saam and Harrer 1999, Harrer and Saam 2000.) It is the intention of these authors to differentiate increasingly between behaviour that is regular and rule obeying, conditioned and rule obeying or sanction avoiding and rule obeying. Modelling norms as both macro- and micro- objects, objective social facts and mental constructs, this research is able to operationalise the insight of Weber in distinguishing between objectively observed behaviour and subjectively intended action.

Castelfranchi, Conte and Paolucci (1998) and Saam and Harrer (1999) investigate the costs of complying with norms since norm-governed systems do not only have advantages but also disadvantages. Under certain circumstances, the normative agents must bear the cost of norms alone while other agents benefit from their presence. Castelfranchi, Conte and Paolucci (1998) introduce communication and normative reputation into the design of their agents. As a result, the spreading of reputations by communication allows normative agents to co-operate without deliberation at the expense of non-normative agents, thereby redistributing the costs of normative action. Referring to sociological theories of deviant behaviour, Saam and Harrer (1999) introduce power into their agent design. Powerful agents decide about the institutionalisation of norms in society as a whole and about the redistribution of wealth. To maintain this social order they regularly have to spend some of their resources. Whenever a powerful agent sanctions a non-normative agent this increases its power and decreases its resources. The resources are replenished depending on an agent's power. As a result, the costs of normative action are redistributed to all agents. However, both studies prove that there are critical values of parameters that determine the volume and scope of redistribution. Under certain circumstances the normative agents bear the cost of norms alone while other agents benefit from their presence.

3.2. NORM PROCESSING

In order to be effective, norms must be processed cognitively. An agent has to recognise that a norm applies and to decide whether to comply with that norm or whether to deviate. Moreover, due to intra- or inter- role conflicts opposing norms may apply to the same agent. In this section, we want to introduce two complementary formalised conceptions of norm processing. The approach of Conte and Castelfranchi (1995b) formalises the recognition of a prescribed norm, as well as alternative ways to adopt norms. Ziegler (1994) elaborates and exemplifies Lindenberg's framing model (1990). He demonstrates how social norms may be processed rationally and in competition with alternatives through the mechanism of framing.

3.2.1. From recognition of prescribed norms to norm adoptation

Conte and Castelfranchi elaborate four alternative ways of moving from recognition of applicable norms to norm adoption. Here, we are only going to describe their central argument. Many more details may be found in their book. According to Conte and Castelfranchi, agents have recognised a prescribed norm if they have normative beliefs. In consequence, agents think that there is an obligation on a given set of agents to do some action:

$$(\text{N-BEL}\ x\ y_i\ a) \overset{def}{=} (\Lambda_{i=1,n}\ (\text{BEL}\ x\ (\text{OUGHT}\ (\text{DOES}\ y_i\ a)))) \qquad (1)$$

x has a normative belief about action a relative to a set of agents y_i if and only if x believes that it is obligatory for y_i to do action a (ibid. 94). For x to believe that a given norm applies, it also needs to believe that it is a member of the class of agents addressed by that norm, a normative belief of pertinence:

$$(\text{P-N-BEL}\ x\ a) \overset{def}{=} (\Lambda_{i=1,n}\ (\text{N-BEL}\ x\ y_i\ a)) \wedge (\text{V}_{k=1,n}\ (\text{BEL}\ x\ (\text{X} = y_k))) \qquad (2)$$

x has a normative belief of pertinence when it has a normative belief relative to a set y_i and an action a, and believes that it is included in y_i (Conte and Castelfranchi 1995b, 99). If x has a normative belief of pertinence, it may adopt the norm and chose action a in an automatic, utilitarian, value driven or Kantian way (*table 2*).

Norm adoption	**Algorithm**
Automatic	IF (P-N-BEL x a) THEN a
Utilitarian	$\forall a \exists p$ [(N-ADOPT x a) $\supset$ ((P-N-BEL x a) $\wedge$ (BEL x ((DONE x a) $\supset$ (OBTAIN x p))))]
Value-driven	$\forall a \exists p$ [(N-ADOPT x a) $\supset$ ((P-N-BEL x a) $\wedge$ (BEL x ((INTEREST y_i (DONE a) p))))]
Kantian	($\Lambda_{x=1,n}$ (N-ADOPT x a_x) with a_x being the set of norm-actions required of x

Table 2: Alternative mechanisms of norm adoption from Conte and Castelfranchi (1995b, 101)

Automatic norm adoption follows a production rule. If x has a normative belief of pertinence, it will perform action a. Agents have neither reasoning powers nor autonomy. In the case of utilitarian norm adoption, the agents are granted greater autonomy. For all norms, x will have the corresponding normative goals if she believes she can get something in return (avoid punishment, obtain approval). Norm adoption is value driven if the agent autonomously shares the end intended by the norm and the belief that the norm achieves that end. According to Conte and Castelfranchi (1995b, 101) Kantian norm adoption is not a rule, but a meta-goal. This means that x observes the whole set of norms facing it as ends in themselves.

There are some parallels between the conception of Conte and Castelfranchi (1995b) and Kohlberg's (1971) reconstruction of the development of human moral consciousness. However, there is also a prominent omission. Conte and Castelfranchi's utilitarian norm adoption corresponds exactly to Kohlberg's first stage of pre-conventional morality (ibid. 164). However, his second stage of conventional morality is completely omitted. Most interestingly, this second stage corresponds to the "sociological" conceptualisation of norms. The possibility that agents may simply adopt a norm because they have internalised it and want to comply with personal expectations and social order is disregarded. At Kohlberg's third and final stage of post-conventional morality, a subject complies with norms because of insight into abstract principles that allow for the foundation and justification of norms. This is the ethical conceptualisation of norms and value driven norm adoption and corresponds to Kohlberg's lower level of post-conventional morality, whereas Kantian norm adoption corresponds to his higher level. In sum, Conte and Castelfranchi have presented an important formalisation of norm processing. From the sociological point of view, however, it has some shortcomings:

(i) Not any norm holds in any context. Instead, norms are context specific (König 1969, 980, Korthals-Beyerlein 1979, 132). An agent has to decide about the relevance of a norm in a certain context. The context has been omitted in the formalisation of norm processing. Ignoring Garfinkel's criticism, this omission can be remedied quite easily. We introduce situation s into equation (1):

$$(\text{N-BEL}\ x\ y_i\ a) \overset{def}{=} (\Lambda_{i=1,n}\ (\text{BEL}\ x\ (\text{OUGHT}\ (\text{DOES}\ y_i\ a\ s)))) \qquad (1b)$$

x has a normative belief about action a relative to a set of agents y_i if and only if, x believes that it is obligatory for y_i to do action a in situation s.

(ii) Kohlberg's second stage of conventional morality which corresponds to the sociological conceptualisation of norms has been omitted. Again, this omission can be remedied quite easily. If x has a normative belief of pertinence, it may adopt the norm and chose action a simply because it has internalised that norm and wants to conform to personal expectations and social order. We shall call this loyal norm adoption.

$$\forall a \exists p\ [(\text{N-ADOPT}\ x\ a) \supset ((\text{P-N-BEL}\ x\ a) \wedge (\text{BEL}\ x\ (\text{MAINTAIN}\ (\text{OUGHT}\ (\text{DOES}\ y_i\ a\ s)))))] \quad (3)$$

For all norms, x will have the corresponding normative goals if it believes that the social order should be maintained.

(iii) The third omission is more fundamental. We are not told when agents resort to automatic, utilitarian, value-driven or Kantian norm adoption. In the following discussion, we shall concentrate on this problem and present Lindenberg's framing model which can be used to discriminate different norm processing procedures.

3.2.2. Selecting norm processing procedures

The key idea of Lindenberg's framing model is that "a situation is framed by a goal (and the relevant goal criterion) in the sense that that goal will select the relevant alternatives and thereby 'define' the situation" (Lindenberg 1990, 743). Norms are conceptualised as goals that are prescribed in certain situations. Norms are processed through the mechanism of framing (Lindenberg 1990, 743). Within rational choice theory, there are different approaches to the formalisation of framing. Most prominent is Lindenberg's discrimination model (Lindenberg 1981). However, in the following discussion, we shall present Ziegler's model (1994) because it is short and demonstrative.

It is Ziegler's intention to demonstrate that there is no basic contrast between rationality and social norms (Ziegler 1994, 151). This is of minor interest to the present discussion. However, Ziegler conceptualises norms as frames and this is most interesting for us. He uses an example, Havel in the tram, a small scene that was reported by Vaclav Havel in his *Letters to Olga* (Havel 1988). It was Steven Lukes who first used this scene to demonstrate rational choice thinking. Havel is about to board the tram:

> "At night, I board the rear car of a tram to go one stop. The car is empty, and since the fare is paid by dropping a crown into a box, not even a conductor is present.... So I have the option of throwing the fare in the box or not: if I don't, no-one will see me, or ever find out: no witnesses will ever be able to testify to my misdemeanor. So I'm faced with a great dilemma, regardless of how much money I happen to have with me: to pay or not to pay." (Havel 1988, 344)

Subsequently, Havel reflects on responsibility. He asks himself why he feels the urge to pay and why the thought of not paying makes him feel guilty. Ziegler formalises this scene in two steps. First, he develops a simple, one-stage subjective expected utility (SEU)-model. In the generalised SEU model, individuals in decision situations choose the alternative that maximizes their subjectively expected utility.

$$SEU_a = p_a * U_a \tag{4}$$

The subjective expected utility *SEU* of an action *a* is equal to the subjective probability *p* that *a* will happen multiplied by the utility *U* of action *a*. Accordingly, Havel has two alternatives:

$$SEU(c) = t - f + g \tag{5}$$

$$SEU(\neg c) = t - \pi(f + s) \tag{6}$$

Havel may comply with the norm and pay the fare (action *c*). He will receive the utility *t* of going one stop by tram. He has to pay the fare *f* and he will have a clear conscience *g*. Havel may also violate the norm and fail to pay the fare (action $\neg c$). He will receive the utility *t* of going one stop by tram. He will have to pay the fare *f* and a fine *s* with probability π. Rational choice theory assumes that if Havel is a rational actor he will comply with the norm if $SEU(c) \geq SEU(\neg c)$, i.e. if $\pi \geq (f - g)/(f + s)$.

Now, there is a critique that argues that individuals often do not consider all subjectively perceived alternatives in their decisions let alone all possible alternatives. How may the relevant alternatives be obtained? Ziegler's answer is framing. The situation is framed by a goal. The goal determines the relevant alternatives and defines the situation. Ziegler develops a second two-stage model. In the first stage, a frame is chosen. In the second stage, the relevant alternatives are evaluated within the chosen frame. Havel may select one of two frames:

$$U_n = SEU(c) = t - f + g \tag{7}$$

$$U_k = max\ \{SEU(c), SEU(\neg c)\} = max\ \{t - f + g, t - \pi(f + s)\} \tag{8}$$

The frame "compliance with norms" *n* considers only one alternative: comply with the norm of paying the fare. The utility of this frame U_n is equal to the subjective expected utility of action *c*. The second frame "calculation" *k* considers normative and deviant action. Here, Havel would act expediently. The utility of this frame U_k results from selecting the maximum utility from actions *c* and $\neg c$. If the frame "compliance with norms" is chosen the norm is adopted automatically because no alternative is considered. If the frame "calculation" is chosen two alternatives are evaluated. But, how is a frame chosen? It is chosen by the same mechanism, by evaluating the subjective expected utility of the frames on the meta-level. Again, Havel has two alternatives:

$$SEU(n) = p * U_n - (1-p) * U_k \tag{9}$$

$$SEU(k) = (1-p) * U_k - p * U_n \tag{10}$$

p indicates the salience (the dominance) of frame *n*. The salience of frame *n* is defined as the strength of the generally agreed belief in the validity of a norm (Ziegler 1994, 158). The salience of frame *k* is given by *1-p* if there is only one alternative frame. Rational choice theory assumes that if Havel is a rational actor he will select the frame that maximises his subjective expected utility. Frame *n* will be chosen if $SEU(n) \geq SEU(k)$, i.e. if $p \geq U_k/(U_n + U_k) = 1/[1 + (U_n/U_k)]$. Ziegler demonstrates that even if frame *k* is chosen, rational actors may decide to comply with the norm. The decision depends on the size of the salience (for details see Ziegler 1994, 155ff).

In sum, compliance with norms is processed rationally and in competition to alternatives through the mechanism of framing. It is this mechanism of framing that could be applied to the selection of norm processing procedures as offered by Conte and Castelfranchi (1995b). The question is when to adopt a norm in an automatic, utilitarian, value driven, loyal or Kantian way. Following Ziegler's model, we distinguish five sub frames of the frame "compliance with norms". These are compliance with norms in an automatic, utilitarian, value-driven, loyal or Kantian way. Then, we have to specify the salience *p* of each sub frame. Depending on the salience, each agent will make a rational choice of its norm processing procedure.

The most prominent critique of this model is the assumption of hyper-rationality in the individuals. Not only are they capable of calculating the subjectively expected utilities of alternative actions but they are also capable of calculating the subjectively expected utilities of alternative frames. While this is a serious issue for sociology, computer science will have no problem with assuming hyper-rationality of co-operative agents.

In this section, we have introduced two complementary formalised conceptions of norms and norm processing. One can think of other formalisations too. From the sociological point of view, these formalisations are some of the simplest we can think of. We are far from knowing which formalisation will be efficient among co-operative agents.

4. Conclusions

Social norms are an important mechanism of coordination among human actors. However, they are in competition with alternative mechanisms. Sociologists have not yet been able to give an empirically validated explanation of this competition. Consequently, there are no complete sociological models of interactivity that may be transferred from sociology to Distributed Artificial Intelligence. Furthermore, social norms are not only solutions to problems of coordination and co-operation. They also generate social problems themselves. DAI will import these problems from sociology unless mechanisms are developed to eliminate them. To develop (or identify) co-operative agents that have the capacity for norm processing one has to advance step by step. Firstly, developing norm concepts for software systems. Secondly, developing conceptions of norm processing. Thirdly, determining the contents of prescribed norms and (finally) developing conceptions that allow norms to change their contents.

Sociology will profit from DAI's research in social norms. We expect that the models of norm processing implemented by DAI will also advance sociological insight into norm processing in human societies. DAI models can be interpreted as social experiments and compared to human societies. For example, sociology may learn from the mechanisms developed by DAI to eliminate the problems that are generated by social norms. At best, sociology will be supported in discovering those mechanisms (and variables) that have remained undiscovered up to now, but are most important because they explain the relation between attitudes and behaviour. In the long run, this could not only prove to be a crucial step in improving the adequacy of theory, but also in increasing the average explained variance of the kind of causal models developed in empirical social research.

References

Bates, J. (1994). The role of emotion in believable agents. In: *Communications of the ACM*, 37, 122-125.

Brenner, W., Zarnekow, R. and Wittig, H. (1998). *Intelligente Softwareagenten: Grundlagen und Anwendungen.* Springer-Verlag, Berlin.

Carley, K. and Newell, A. (1994). The Nature of the Social Agent. In: *Journal of Mathematical Sociology*, 19, 221-262.

Castelfranchi, C., Conte, R. and Paolucci, M. (1998). Normative reputation and the costs of compliance. In: *JASSS - Journal of Artificial Societies and Social Simulation* 1(3), <http://www.soc.surrey.ac.uk/JASSS/1/3/3.html>.

Castelfranchi, C. and Werner, E. (1994). The MAAMAW spirit and this book. In: Castelfranchi, C. and Werner, E. (eds.) *Artificial Social Systems.* Springer-Verlag, Berlin, ix-xiii.

Coleman, J.S. (1986). Social Structure and the Emergence of Norms Among Rational Actors. In: Diekmann, A. and Mitter, P. (eds.). *Paradoxical Effects of Social Behavior: Essays in Honor of Anatol Rapoport.* Physica-Verlag, Heidelberg, 55-83.

Coleman, J.S. (1986/87). The Emergence of Norms in Varying Social Structures. In: *Angewandte Sozialforschung*, 14, 17-30.

Coleman, J.S. (1990a). *Foundations of social theory.* The Belknap Press of Harvard University Press, Cambridge, MA.

Coleman, J.S. (1990b). The Emergence of Norms. In: Hechter, M., Opp, K.-D. and Wippler, R. (eds.). *Social Institutions: Their Emergence, Maintenance and Effects.* de Gruyter, Berlin, 35-59.

Conte, R. and Castelfranchi, C. (1995a). Understanding the functions of norms in social groups through simulation. In: Gilbert, N. and Conte, R. (eds.). *Artificial Societies: The Computer Simulation of Social Life.* UCL Press, London, 252-267.

Conte, R. and Castelfranchi, C. (1995b). *Cognitive and social action.* UCL Press, London.

Dahrendorf, R. (1962). On the Origin of Social Inequality. In: Laslett, P. and Runciman, W. G. (eds.). *Philosophy, Politics and Society.* Basil Blackwell, Oxford.

Durkheim, E. (1966) [1893]. *The division of labour in society.* third edition, Free Press, New York.

Durkheim, E. (1965) [1895]. *Regeln der soziologischen Methode*. second edition, Luchterhand, Neuwied.
Elster, J. (1989a). *The cement of society*. Cambridge University Press, Cambridge.
Elster, J. (1989b). Social Norms and Economic Theory. In: *Journal of Economic Perspectives*, 3, 99-117.
Elster, J. (1991). Rationality and Social Norms. In: *Europäisches Archiv für Soziologie*, 32, 109-129.
Esser, H. (1993). *Soziologie: Allgemeine Grundlagen*. Campus, Frankfurt.
Garfinkel, H. (1967). *Studies in Ethnomethodology*. Prentice-Hall, Englewood Cliffs, NJ.
Geiger, T. (1962). *Arbeiten zur Soziologie*. Luchterhand, Neuwied.
Harrer, A. and Saam N.J. (2000). Social norms, inequality, and functional change in artificial societies - Some more results. In: Urban, C. (ed.). *Workshop 2000: Agent-Based Simulation*. Society for Computer Simulation International, Erlangen, 13-18.
Hart, H.L.A. (1961). *The concept of law*. Oxford University Press, Oxford.
Havel, V. (1988). *Letters to Olga: June 1979 to September 1982*. translated by Paul Wilson, Faber and Faber, London.
König, R. (1969). Soziale Normen. In: Bernsdorf, W. (ed.). *Wörterbuch der Soziologie*. second edition, Enke, Stuttgart, 978-983.
Kohlberg, L. (1969). Stage and Sequence: The cognitive development approach to socialization. In: Goslin, D.A. (ed.). *Handbook of socialization theory and research*. Rand McNelly College Publishing Company, Chicago, IL.
Kohlberg, L. (1971). From is to ought. In: Mischel, T. (ed.). *Cognitive development and epistemology*. Academic Press, New York.
Korthals-Beyerlein, G. (1979). *Soziale Normen. Begriffliche Explikation und Grundlagen empirischer Erfassung*. Fink, München.
Lewis, D. (1969). *Convention*. Harvard University Press, Cambridge, MA.
Lindenberg, S. (1981). Rational repetitive choice: The Discrimination Model versus the Camilleri-Berger Model. In: *Social Psychology Quarterly*, 44, 312-330.
Lindenberg, S. (1985). An assessment of the new political economy: Its potential for the social sciences and for sociology in particular. In: *Sociological Theory*, 3, 99-114.
Lindenberg, S. (1990). Homo socio-oeconomicus: The emergence of a general model of man in the social sciences. In: *Journal of Institutional and Theoretical Ecomonics*, 146, 727-748.
Lukes, S. (1991). *The Rationality of Norms*. Unpublished manuscript, European Social Theory Conference on "Rational Choice Perspectives in Sociological Theory and Analysis", European University Institute, San Domenico di Fiesole (Firenze), Italy, June.
Malsch, T. (1997). Die Provokation der "Artificial Societies": Warum die Soziologie sich mit den Sozialmetaphern der Verteilten Künstlichen Intelligenz beschäftigen sollte. In: *Zeitschrift für Soziologie*, 26, 3-21.
Marx, K. (1936). *Capital, a critique of political economy: The process of capitalist production*, translated from the third German edition by S. Moore and E. Aveling, edited by F. Engels, revised and amplified according to the fourth German edition by E. Untermann, The Modern Library, New York.
Moffat, D. and Frijda, N.H. (1995). Where there's a Will there's an Agent. In: Wooldridge, M.J. et al. (eds.). *Intelligent Agents*. Springer-Verlag, Berlin, 245-260.
Müller, H.J., Malsch, T. and Schulz-Schaeffer I. (1998). Socionics: Introduction and Potential. In: *JASSS - Journal of Artificial Societies and Social Simulation* 1(3), <http://www.soc.surrey.ac.uk/JASSS/1/3/5.html>.
Nonner-Winkler, G. (1984). Normen. In: Kerber, H. and Schmieder, A. (eds.). *Handbuch Soziologie: Zur Theorie und Praxis sozialer Beziehungen*. Reinbek, Rowohlt, 398-408.
Parsons, T. (1951). *The Social System*. Routledge and Kegan Paul, London.
Piaget, J. (1983). *The moral judgement of the child*. Penguin Books, Harmondsworth.
Saam, N.J. and. Harrer, A. (1999): Simulating Norms, Social Inequality, and Functional Change in Artificial Societies. In: *JASSS - Journal of Artificial Societies and Social Simulation* 2(1), <http://www.soc.surrey.ac.uk/JASSS/2/1/2.html>.

Schelling, T. C. (1960). *The strategy of conflict*. Oxford University Press, Oxford.
Sherif, M. (1967). Formation of social norms: The experimental paradigm. In: Sherif, M. (1967). *Social Interaction*. Aldine Publishing Company, Chicago, IL.
Shoham, Y. (1993). Agent-oriented programming. In: *Artificial Intelligence*, 60, 51-92.
Shoham, Y. and Tenneholtz, M. (1992a). On the synthesis of useful social laws for artificial agent societies (preliminary report). In: *Proceedings of the AAAI Conference*, 276-281.
Shoham, Y. and Tenneholtz, M. (1992b). Emergent conventions in multi agent systems: Initial experimental results and observations. In: *Proceedings of the Third International Conference on KR&R*. Cambridge, MA, 225-232.
Ullman-Margalit, E. (1977). *The emergence of norms*. Clarendon Press, Oxford.
Walker, A. and Wooldridge, M.J. (1995). Understanding the emergence of conventions in multi-agent systems. In: *Proceedings of ICMAS (International Joint Conference on Multi Agent Systems)*, San Francisco.
Weber, M. (1960). Soziologische Grundbegriffe. In: Weber, M. *Wirtschaft und Gesellschaft*. Mohr, Tübingen.
Wooldridge, M.J. and Jennings, N.R. (1995). Agent Theories, Architectures, and Languages: A Survey. In: Wooldridge, M.J. et al. (eds.). *Intelligent Agents*. Springer-Verlag, Berlin, 1-39.
Ziegler, R. (1994). Normen, Rational Choice und Netzwerkanalyse. In: Derlien, H.-U. et al. (eds.). *Systemrationalität und Partialinteresse*. Nomos, Baden-Baden.

CAN AGENT-BASED MODELLING REALLY BE USEFUL?

JIM E. DORAN
Department of Computer Science
University of Essex
Colchester, United Kingdom

Abstract

Agent-based modelling on a computer appears to have a special role to play in the development of social science and the formulation of social policy. It offers a means of discovering general and applicable social theory, and grounding it in precise assumptions and derivations, whilst addressing those elements of individual cognition that are central to human society. However, there are important questions to be asked and difficulties to be overcome in achieving this potential. What differentiates agent-based modelling from traditional computer modelling? What different types of agent-based models are there, and what are the structural relationships between them (if any)? Which model types should be used in which circumstances? If it is appropriate to use a complex model, for example one incorporating "deliberative" agents, how can it be validated? If it can only be validated in general terms, does this mean that we are *forced* into a "theory building" mode in which the focus of the investigation lies in the model's properties? If so, what types of parameter space may a complex model have? How best can very large parameter spaces be explored? Some of these questions are here addressed and are illustrated by reference to recent agent-based models for the environment. A particular application is then considered in some detail: agent-based modelling of intervention strategies for integrated ecosystem management, especially management of the Fraser River watershed in British Columbia.

1. Introduction

Agent-based modelling is attracting significant attention from social and environmental scientists (e.g. Gilbert and Troitzsch 1999, chapter 8; Carpenter et al. 1999; Gilbert 2000; Moss 2001). Indeed it has been proposed that this type of computer modelling can be a useful source of input to policy making in those problem contexts in which there is a combination of social and natural processes (e.g. Bousquet *et al* 1994). Major projects inspired by this prospect are now under way (e.g. FIRMA 2000).

But is optimism justified? Agent-based modelling encounters major methodological difficulties. These could prove insurmountable. To evaluate these difficulties and their potential resolution we consider recent examples of

N.J. Saam and B. Schmidt (eds.)
Cooperative Agents. Applications in the Social Sciences, 57-81

the use of agent-based modelling and their successes and limitations. We also look at a hard but particularly important problem that agent-based modelling should in principle be able to address – integrated watershed management.

2. Agent-Based Modelling

Agent-based modelling is essentially computer simulation, but using complex models that include *agents*. In this context, agents are software entities that are "autonomous" loci of decision making. Agents sense, decide and act. *Reactive* agents choose actions directly by reference to their sensed circumstances. *Deliberative* agents are those which, for example, reflect upon alternative courses of action, and select one of them for execution. That is, they plan. *Adaptive* agents change their behaviour in the light of changing circumstances. *Social* agents communicate and co-operate with other agents.

Agent-based modelling draws upon rapidly developing *agent technology* (Jennings and Wooldridge 1998) which is a mixture of artificial intelligence and other computer science techniques, notably object-oriented programming. A range of alternative agent software design architectures have been developed, including architectures based upon AI planning systems, upon simple or "fuzzy" rules, upon artificial neural networks, and hybrid architectures. Possible relationships between agents include co-operation and negotiation, subordination, and competition. Experimental inter-agent communication languages have been developed notably KQML and ACL.

In agent-based models of components of human societies, agents are often interpreted as in correspondence with individual humans, with small human groups, or with organisations.

2.1. THE PROMISE

At its most successful, modelling (not necessarily agent-based modelling) promises major insight into the system modelled, the target system, leading to scientific advance, and useful prediction of the target system's behaviour leading to practical policy guidance. Even when, as is all too often the case, these primary objectives are at best only partially achieved there may be other "soft" benefits of a modelling study. For example a study may encourage the collection of data and the recognition of existing gaps in knowledge. It may enhance communication between researchers and better focus their discussions (Downing, Moss and Pahl-Wostl 2001). It may serve as a useful educational tool. Minimally, a modelling study will illustrate the concepts that inform and guide it. And quite often an attempt at modelling bring forward one or more interesting technical problems, which then embark upon a life of their own.

However, the further and particular promise of *agent-based* modelling is that it enables us to work with computer models that explicitly get to grips with the interaction between individual and social cognition; with the interaction between the real world and the collectively, that is socially, conceived world. Few would seriously deny that the beliefs and goals of individuals often influence and are influenced by the structure and dynamics of human society. Agent-based models with non-trivial agents within them enable such matters to be addressed.

3. Three Recent Examples Of Agent-Based Modelling

We briefly describe three recently reported examples of agent-based modelling projects, that are representative of agent-based modelling in general and also relate to our theme of environmental management (see also Rouchier *et al.* 2001).

Carpenter, Brock, and Hanson (1999) report a set of three agent-based models of a typical lake subject to phosphorous pollution and eutrophication. Each of the models also captures something of the physical and agricultural environment of the lake. The models are expressed in terms of difference equations that specify the lake, runoff etc., and seek to capture self-interested, rational decision making by framers and others given their partial and possibly incorrect information. The three different models address different types of management regime:

- management by *market manager* (assuming a market in latest pollution data),
- management by *governing board* (an elected board, whose members differ in their views of "nature" and how to manage it), and
- management by *land manager* (who sets targets, regulations and incentives).

Because no specific lake is targeted, the conclusions reached are presented as indicative only. Significantly, the experimenters observe "irregular oscillations among ecosystem states and patterns of agent behaviour" for a wide range of scenarios. One particular conclusion they reach is that there is a "need for experimentation, learning and adaptation" by real managers (see later, *adaptive management*).

Moss (1999) reports an agent-based model of a specific organisation (North West Water, UK). The model is directed to enhancing understanding of the company's management of critical incidents, and was developed after detailed interviews with company employees. The agents are based upon theories of cognition drawn from cognitive science and are, in a limited way, capable of selecting between simple cognitive rules (which are confusingly called "models" in the published account of the work) which they use to characterise and respond to a sequence of events. The agents are also adaptive in that they adjust the weight that they give to individual cognitive rules. A particular result of the study is to discover the importance of transferring information between successive shift managers.

The agent-based model of Kohler *et al* (2000) is of a slightly different type. These authors have used their model to simulate the development of a settlement pattern from AD 900 to 1300 in a particular region of south western Colorado, USA. The agents in the model are rule-based and correspond to households. They take decisions about relocation and crop planting based upon such factors as local soil conditions, water resources, and recent weather. The simulated individual people in the households take no decisions as such, but are born, reproduce and die. From time to time these events lead a household to generate a new one to be located elsewhere. Current work is enabling households to engage in product exchange as a medium term survival strategy. The output of the model, for alternative combinations of parameter settings, is being compared with a detailed archaeological reconstruction of the actual settlement pattern as it developed in prehistory. Unlike the two previously described studies this work is not directed, even in the longer term, to improvement of any management practice. Rather, it is solely targeted at enhanced scientific understanding.

4. Problems Of Agent-Based Modelling

The great increase in complexity in a computer-based model that follows from the use of deliberative or even reactive software agents within it brings substantial methodological problems in its train. We now discuss a number of major problems typically encountered in practical agent-based modelling work, with illustrative reference to the projects described in the previous section.

4.1. SKILLS AND TIME REQUIREMENT

The task of designing, constructing and experimenting with an agent-based model is a relatively demanding one. Substantial specialist expertise is required in the technical domain of the model (e.g. in artificial intelligence, agent technology, or difference equations) in order to make and carry through the detailed choices required. The time to create and debug the software implementation is likely to be substantial. The implication is that a multidisciplinary research team will be needed for the work that brings together both the technical expertise required for the creation and handling of the actual model, and the social science expertise needed to analyse the target system. This need is amply illustrated in the projects described above, most obviously in that of Kohler *et al.*

There is, unfortunately, a tendency to "cut back" on the required modelling expertise on the mistaken assumption that it is subordinate to the social science content. This is potentially disastrous. It can easily lead to shallow and ill-conceived models that merely act, at best, as a sounding board for a particular social science perspective.

4.2. WHICH TYPE OF MODEL?

There are alternative types of model even within an agent-based formulation. The term "agent" itself is somewhat ambiguous, and a range of different agent architectures have been developed (see earlier, section 2). For example, the detailed technical content and connotations of the agents in the various models described earlier are quite different. Carpenter *et al* put the emphasis on rationality and cost-benefit analysis, and design their agents accordingly. Moss adopts an agent design derived from cognitive science. Kohler *et al,* are content with a "condition-action" rule set for each agent, representing a household. There is little guidance available in the literature on how to choose between such agent design alternatives in any reasoned way and investigators are obliged to rely largely on their disciplinary background whether or not it is fully appropriate. Even more seriously, it sometimes seems to be overlooked that we (still) do not know how to program agents with cognitive abilities at anything like human level. Only agent-based models in which agents have *very limited* cognitive abilities are feasible, however useful models incorporating more "intelligent" agents might be.

4.3. WHAT LEVEL OF ABSTRACTION?

Selecting agent type is an aspect of the more general problem of selecting for a model a level or combination of levels of abstraction. It is never possible to construct a model that reflects the target system in comprehensive detail. Always the level of aggregation and abstraction (the "grain of analysis" as it is sometimes called) must be chosen, with the aim of achieving tractability without loss of significant model function. For example, even when a choice has been made of agent type it remains to be decided which elements of the target system are and are not to be individually expressed by agents. The answer may be many or few. Thus the Kohler *et al* study models individual households but not the cognition of individual people. The Carpenter *et al* models do not distinguish individual farmers and their families nor represent individual fish in the lake.

As with choice of model type, there seems to be rather little guidance available on how to choose the level of abstraction in a reasoned way. A very common response is to identify software agents with individual people or with clear-cut organisations or groups of people. It is very rare for any subtler notion of an agent correspondence to be considered.

A promising approach that has occasionally been explored, and which needs further work, is to define a lattice of models with levels in the lattice corresponding to levels of abstraction. Then criteria may be specified and algorithms defined which search over the lattice to locate an appropriate model for some particular modelling task. A basic requirement is that a model should be the simplest which will address usefully the questions to be asked of it. Lee and Fishwick (1997), for example, have recently developed and discussed a dynamic model abstraction scheme and Fishwick, Sanderson and Wolff (1998) have discussed its application to ecosystem management.

4.4. SEARCHING A MASSIVE PARAMETER SPACE

The complexity of the cognitive structure and processes of non-trivial agents implies a wide range of adjustable parameters and a correspondingly massive exogenous parameter space for an agent-based model. This is evident in all three of the projects described earlier. Furthermore, there will be many detailed structural choices. For example, agents that are to construct and execute plans of action, or which are to be adaptive, can be engineered to have these necessarily complex properties in many different ways and to many different degrees. How do we decide which model variants to investigate? How do we find where in the potentially enormous space of parameter settings and structural choices are the significant and interesting phenomena (if anywhere)? The likely presence of a degree of non-determinism in the model makes the task all the more daunting.

At present, this challenge is commonly avoided rather than met, with experimenters tending to assume that they know in advance which (few) adjustable parameters are of importance and need to be experimentally varied, and holding the remainder fixed. Arbitrary truncation of the space, and simple methods of systematic or random sampling, are clearly insufficient. There is a need for techniques and tools of experimentation incorporating heuristic but effective search processes, for example *hill climbing*, and methods of sub-region characterisation (e.g. Hales, 2001; and see Teran, Edmunds and Wallis, 2001). Furthermore, we must keep in mind the possibility of discontinuities or singularities in the parameter space, and the possibility of chaotic model behaviour. There may well be points or regions in which a model's behaviour is effectively without regularity.

A related question is the sensitivity of a model's behaviour to details of its computer implementation. A number of authors have suggested that the published results of modelling studies can sometimes reflect low level and insignificant implementation decisions rather than dynamic properties of the model of real scientific interest. An important example is how to simulate concurrent activity on a single processor machine, where different implementation choices can have very different consequences (see e.g. Lawson and Park 2000, Axtell 2001). There is little doubt that we still have much to learn here.

4.5. THE PROBLEM OF VALIDATION

Finally, and perhaps most important of all, there is the problem of model validation, that is, how to establish that a model does indeed reliably and usefully correspond to reality. Traditionally, the requirement has been that the model should be "checked out" against the target system in sufficient detail to establish confidence in the reliability of any findings made using it. This checking might either be at the level of the model's individual components, or at the level of its behaviour as a whole. But validating an agent-based model is particularly difficult because of its complexity of structure, especially cognitive structure. In fact, it becomes near impossible. For example, it is not difficult to dispute the detailed formulation of the difference equations that constitute the models of lake pollution described earlier, and the Kohler *et al* model of household decision making is clearly quite unreal as regards the actual process of decision making.

One reaction to this problem is simply to do the best one can given available knowledge (as Kohler *et al* do), and another is to argue for some form of general validity for the agent structures (cf. the Moss work). Or validation may be entirely in terms of a demonstrated match between a sample of the model's behaviour and the real behaviour of the target system. Whichever of these roads is followed, the study's conclusions are likely to be limited to broad generalisations (Doran 1997).

Alternatively, we may discard the notion of modelling some specific target system (as do Carpenter *et al*) and instead focus on the discovery of social theory. In this case the emphasis is placed on the use of the model for reliable and objective derivation of the consequences of certain structural assumptions and certain parameter settings. Relevance is claimed to some class of social system rather than to any specific and actual system (see, for example, Epstein and Axtell 1996, Doran 1997).

4.6. SUMMARY REMARKS ON THE METHODOLOGICAL PROBLEMS

It is apparent that the various methodological problems associated with agent-based modelling are substantial and of great practical significance. Their impact in current work is clear. In the past, their cumulative effect has been to limit the effectiveness of agent-based modelling to little more than illustration of what *might* be possible and of how that potential *might* be realised. In all three of the particular studies considered earlier, the actual conclusions drawn have been vague, general and tentative and have added little to common-sense. Does this suggest that agent-based modelling can never be of practical importance? To explore this question further, we now discuss modelling a challenging example of ecosystem management.

5. Ecosystem Management

Environmental concerns have become ubiquitous as global human population increases along with its consumption of natural resources. Explicit and large-scale management of natural resources -- agriculture areas, forests, fishing stocks -- has been forced upon us as we attempt to solve ever-worsening problems of resource decline and failure. Notions of *integrated* management, covering all aspects of an ecosystem, and of *sustainability*, seeking to manage natural resources in ways that secure the future, have become prominent. The precise meaning of *sustainable* is somewhat elusive. A well-known, and cautiously worded, definition of sustainable development is "development that meets the needs of the present without compromising the ability of future generations to meet their own needs" (WCED 1987, 43). Alternatively: "Sustainability means living within the constraints imposed by a finite set of global resources while satisfying the reasonable social and material aspirations of most of the world's citizens." (Healey 1998a). Many would argue that any concept of "sustainable growth" (rather than "sustainable development") is a dangerous oxymoron.

Over the past decade a widespread response to perceived past failures has been to reject centralised, "top-down" and "bureaucratic" management, in favour of *decentralisation* and a much more co-operative and community centred form of management. Since management and its failures impacts peoples' lives in profound ways, social and political reverberations are inevitable. Indeed, certain styles of government and multinational commerce and industry have become major targets of criticism, hostility and even violence. However, it sometimes seems that this move towards decentralised management is at root no more than an emotional and ideological reaction in the face of a global problem that human society is not currently equipped to address.

5.1. INTEGRATED WATERSHED MANAGEMENT

Integrated river watershed management is a category of ecosystem management that is of obvious importance and complexity. Potentially conflicting facet and requirements within a watershed are:

- water supply (for domestic and agricultural uses)
- pollution control
- fisheries management
- flood control
- hydropower production
- navigation and wetlands management
- recreation provision

Biophysical investigation addresses such crucially important matters as water flows, fish population dynamics, types and levels of pollution, and the impact of control procedures. But beyond problems at the biophysical level are human social problems and these are sometimes insufficiently considered.

5.2. STAKEHOLDERS AND WATERSHED SOCIETY

Human society in a watershed typically comprises a multiplicity of organisational *stakeholders*, each pursuing its own interests, with conflicting goal sets, and each with only limited awareness and knowledge of the others. There may well be sharp political and cultural differences. Put in general terms, the task at all scales in a watershed is to execute a strategic program of natural resource utilisation and conservation which balances sectional interests and which is sustainable. To achieve sustainable balance is not easy, and is impossible without restraint and compromise. Again expressed in general terms, a number of particular sources of difficulty may be listed. These include the complex structure of the environmental subsystems, the private agendas of the particular stakeholders and their conflicting norms and expectations, stakeholders' lack of information, and failures of inter-stakeholder communication.

Unfortunately it is rare that a co-operation regime that is beneficial overall will always be to the benefit of every stakeholder. Very often a management strategy that serves the interests of a majority of the population in the long-term will significantly harm the interests of minorities. If this is so, then there is obvious potential for social conflict.

5.3. THE FRASER RIVER WATERSHED

An important particular case of watershed management is that of the Fraser River in British Columbia, Canada (but see also, for example, the controversy surrounding the Columbia River in the USA and the Snowy River in Australia). The management task for the Fraser may be viewed on any of three scales: the entire Fraser watershed; the Lower Fraser Basin and separately, the Upper Fraser; or, recursively, any of the sub-watersheds of the Fraser. At each scale there are particular characteristics and problems as well as the problems common to all three scales.

The Fraser River watershed as a whole is large and diverse ranging from conurbation (Greater Vancouver) to near wilderness. Logging is a major industry, and the salmon populations are of major importance but in serious decline. Over the past decade a major attempt at integrated management of the watershed at this scale has been made, but this has encountered many social and political difficulties and as yet has had limited success (Dorcey 1997; Marshall 1998). A particular problem is the complexity of governance. Federal, Provincial, Municipal and First Nations governments are all stakeholders in the watershed. The aboriginal or "First Nation" dimension adds further distinctive complexity as aboriginal culture becomes influential again and, for example, land rights are (re-)negotiated.

The Lower Fraser Basin (Healey 1999) has experienced a population explosion over the past century, which has resulted in de-forestation, agriculture, industrialisation and conurbation, and rising levels of pollution. There is now a major immigrant population, a majority of Chinese race, whose cultural norms and expectations do not fully align either with those of European or of First Nation origin. This ethnic diversity further complicates environmental issues and policy.

The particular problem of the smaller urban sub-watersheds, for example, the Brunette River watershed (Hall *et al.* 2000) is that any attempt at integrated management must grapple with a range of powerful stakeholders, including any or all of the four orders of government listed earlier, and major corporations. Achieving a coherent watershed management plan (and financing and then implementing it) becomes a delicate matter of initiating and bringing to fruition negotiation between the representatives of these powerful stakeholders, with local community input where possible.

6. Agent-Based Modelling And Watershed Management

From the Fraser River example we discover that watershed integrated management problems are primarily to do with people and organisations and the attitudes that they adopt and that determine their behaviour. It therefore seems that if agent-based modelling is to be used to address watershed integrated management problems, models must address the social and human issues of organisation, conflicting belief systems, partial knowledge, imperfect decision-making, fallible communication, negotiation and the establishment of co-operation. Indeed, it seems that we must focus on *social intervention strategies*. This insight can be made more precise as follows.

6.1. LEVELS OF FORMULATION

We distinguish *three levels*[1] at which the problem of integrated watershed management might in principle be precisely formulated in agent (or other) terms:

- The level (I) of action -- the physical problem of data communication, stakeholder action, and action co-ordination.
- The level (II) of management -- the design of a management regime able to devise and implement an effective solution to the level I problem.

[1] These three levels should not be confused with the three possible scales of consideration of the Fraser River watershed described earlier.

- The level (III) of intervention -- design and implementation of an intervention policy that can bring into existence an effective solution to the level II problem.

As suggested above, it is clear from the Fraser River experience (Dorcey 1997) that the level I problem, although very difficult in practice (and the problem containing the hard science), is arguably not that of the greatest importance. Nor indeed is it the level II social and management structure design problem where the major difficulty lies. The key task is the "people problem" at level III, the task of intervening to reconcile conflicting interests and to engender co-operation. Intervention policies are needed which “move” a discordant "Tower of Babel" society that speaks with many voices and is not even fully aware that it has a problem of achieving integrated management for sustainability, to one that actually co-ordinates to solve the problem. Thus the emphasis moves towards communication, intervention and persuasion. Agent-based modelling is of potential value just because it can plausibly extending computer modelling into this arena.

7. A Micro-Model Of Intervention - The Devil's In The Detail

We now introduce and extend a well-known test problem of artificial intelligence in order to examine rather more precisely the notion of *intervening* in a multi-agent system. The investigation will shed a little light on possible intervention strategies for an agent-based model of ecosystem management.

7.1. THE CLASSICAL "MONKEY AND BANANAS" (MANDB) PROBLEM

The classical AI *Monkey and Bananas* problem (e.g. Russell and Norvig 1995, 366) is typically stated somewhat as follows:

> There is a monkey in a room with some bananas hanging out of its reach from the ceiling. A box is in the room. The monkey and the box are at the same height on the floor of the room but at different locations. If the monkey is on the box it will be at the same height as the bananas. The actions available to the monkey are: *Go* from one location to another, *Push* an object from one location to another, *Climb* onto an object, *Grasp* an object, and *Eat* any bananas it is grasping. The action *Grasp* requires the object grasped to be at the same height and place as the monkey.

The problem is "What should the monkey do?" or, more precisely, to so design the monkey (an agent) that it gets to eat the bananas. Normally this is taken to be a matter of the monkey starting with the goal of eating bananas, and then planning how to obtain and eat the bananas and then executing its plan.

This problem is rather easy for humans, but less so for real monkeys. Its interest lies in the precise details of how the problem scenario, including the monkey's available actions, is to be formally specified (there are many possible alternatives), and what exactly is the plan generation and execution process to be followed (again there are many alternatives). The main elements of the problem specification are:

- formulae which describe the initial state of the world
- an initial goal for the monkey
- specifications of the actions available with their preconditions and effects (e.g. in STRIPS format)

The *solution* to the problem is typically to design the monkey to employ a *backward chaining* planning algorithm, followed by forward plan execution. For details of a particular specification of the classical MandB problem see Nilsson (1971, 35-37).

7.2. EXTENDING THE MONKEY AND BANANAS PROBLEM

If the MandB problem is to be related to sustainable resource management, it clearly needs elaboration. Let us consider the following more complex scenario:

> There are two monkeys in the room, one box, and 100 bananas hanging from the ceiling.
>
> Either monkey standing on the box (in the appropriate location) can grasp 60 of the bananas, but only 60. The remaining 40 of the bananas cannot be grasped from the box (by either monkey).
>
> Either monkey is physically able to climb upon the other's shoulders
>
> Either monkey standing upon the other's shoulders, when the other is on the box, can reach *all* the bananas (but can also grasp less than 100 if it so "chooses")
>
> We assume that the bananas are replaced with fresh ones every day *provided that at least 10 of the bananas remain from the previous day.* If ever all bananas are removed on a day, then *never again* are bananas supplied (and presumably the monkeys "die").

The last condition means that, from the viewpoint of the monkeys, taking all the bananas is attractive in the short term but disastrous in the longer term.

We might prefer to see the monkeys adopt the following "equitable" and "sustainable" solution course of action.

> Every day, one monkey is to stand upon the shoulders of the other, to take just 90 of the bananas, and then equally to share the bananas taken with the other. Thus each monkey gets to eat 45 bananas each day, indefinitely.

Notice that use of the box rather than co-operation makes only 60 of the bananas available, and taking all the bananas (*collective greed*) is disastrous.

7.3. THE EXTENDED SPECIFICATION IN MORE DETAIL

In the elaborated scenario there must be distinguished (i) an actual state of the room and (ii) for *each* of the monkeys a *belief state* comprising:

- a fixed goal for the monkey (to eat bananas -- as many as it can).
- beliefs which describe its (sensed) awareness of the state of the room. These beliefs may be incomplete or even in error as compared with the actual state of the room, and are continuously updated.
- specifications of the actions available to the monkey (which again may be incomplete or in error) with their preconditions and effects.

In such specifications the beliefs are commonly expressed in the language of predicate logic. As regards actual possible actions, assume that the grasping of different percentages of bananas is represented by a set of different actions e.g. *Grasp-60*. Assume also that one monkey can (physically) *Climb* upon another. *Push* still applies only to the box. We also require a new set of actions *Give* e.g. *Give-50*.

We assume that both monkeys possess and use the same planning algorithm, and that this algorithm is capable of generating a *joint* plan *if the necessary information exists in the monkey's current belief state.* This means that one monkey may formulate a plan in which an action of the other is assumed. The other's action is then an instance of a possible *event* for the monkey doing the planning i.e. something that may happen without action by the monkey itself provided certain preconditions are met. Of course, incorporating such an event into the plan does not of itself mean that the action of the other is in any way guaranteed to occur.

7.4. THREE SCENARIO VARIANTS AND THEIR OUTCOMES

We can now make more detailed assumptions about the extended MandB scenario, here in just three alternative ways, and consider what the outcome will be in each case. *It is important to keep in mind in what follows that the monkeys (agents) only have the limited abilities that have been specifically assigned to them.* Everything mentioned in these three scenario variants is feasible in software -- e.g. see Doran and Palmer (1995).

Scenario Variant I. Firstly assume that neither monkey can sense the other, so that none of its beliefs refer to the other. What will happen when the scenario is run?

> Each monkey can only plan for itself, as before, and there will be, in effect, a "race" to push the box and one monkey will win it. The winning monkey will then take and consume 60 bananas. The loser will consume nothing. This race will happen daily, not necessarily always with the same outcome.

Scenario Variant II. Assume that each monkey does have awareness of the other (but not the other's mental state). What will happen when this scenario is run?

> The ultimate outcome will be much as before. A monkey can create joint plans (in its own interest, taking 100 bananas) but has no useful way to interact with the other. In particular there is no communication between the monkeys. Therefore there is no way in which one monkey can "move" the other to a particular location to stand on its shoulders. There may be some delay as the monkeys each formulate joint plans, which they then "decide" (how?) that they cannot execute, or actually fail to execute.

Scenario Variant III. Finally, let us go one step further and provide each monkey with a *Request* action. This action, when it is performed, sends a message to the other monkey which, in effect, promises to the other that a certain quantity of bananas (say 50) will be provided "later" if a specified action is performed by the other right away. (We assume without elaboration that a monkey is able to receive, process and utilise such messages appropriately, although this is non-trivial to implement in software). Now what will happen?

> Both monkeys will create and evaluate joint plans in their own interest i.e. each will create a plan to use the other to obtain 100 bananas for itself. Each plan will include a *Request* to the other to move to the location below the bananas in return for 50 bananas. However, each monkey will also, as before, consider a plan in which it acts independently (pushing the box into position, and climbing upon it etc) with an expectation of obtaining 60 bananas. Each will prefer this individual plan, which appears to yield more bananas, so that again there will be a race.

Of course, there is much more to be considered by way of enlarging a monkey's repertoire of abilities. Will a monkey receiving a request accept it -- presumably not as an individual plan "pays more"? Will a monkey that has issued a request keep its promise to supply bananas "later"? What if a monkey has the ability to consider *not* keeping a promise? What if a monkey has the ability to consider offering less than 50 bananas in return for a co-operative action? What if a monkey has the ability to assess the probability that a plan will succeed rather than fail? What if a monkey has the ability to calculate the longer-term implications of its immediate actions?

7.5. TYPES OF INTERVENTION

We now turn to the issue of *intervention* in the extended MandB scenario. Intervention here refers to an experimenter outside the scenario sending messages to either or both of the monkeys which to lead them to co-operate in our preferred way. We may identify the following categories of messages:

- Statements of asserted *fact* -- e.g. there is a monkey at location X
- Specifications of *actions* or *events* (in the sense indicated earlier)
- Statements of *goals*

Assume that monkeys always *reject* statements of goals, but *accept* statements of fact and new action/event specifications provided that these are not contradicted by immediate sensing. This means that intervention is essentially a matter of supplying information to either or both monkeys, about the state of the world, or actions or events.

The provision of *accurate* information certainly does not guarantee the outcome we want, as has been exhibited in the scenario variants explored above. However, there is also the possibility of supplying the monkeys with *inaccurate* information. For example, suppose that we inform monkey A (alone) of the following event -- any monkey who stands beneath the bananas will in due course receive 120 bananas (a lie!). The effect will be that monkey A will indeed stand beneath the bananas and monkey B will be able to climb upon it to collect 100 bananas. We also inform monkey B of a different event -- that by giving monkey A 50 bananas, monkey B will itself acquire 200 of the bananas (another lie!). The effect of this intervention looks to be that both monkeys end up with 50 of the bananas.

Our intervention needs to be modified so that only 90 of the bananas are taken and then equally shared, so ensuring sustainability. However, this is not entirely straightforward. There is no immediate way to stop a monkey taking 100 of the bananas when it is in a position to take 90. The way that we have formulated intervention means that we cannot directly suppress the *Grasp-100* action, nor can we have the monkey believe that there are only 90 bananas to take, since this contradicts sensing. One possibility is to further extend the scenario slightly, by making the top goal of a monkey be to survive (i.e. not to die), retaining eating bananas as a sub-goal. A monkey may then be informed that it will die if it takes 100 bananas, which will discourage it from doing so.

7.6. DISCUSSION

This little investigation using the extended MandB problem reveals the complexity of the scenario once it is pitched at the AI technical level -- as it must be to be relevant to agent-based modelling on a computer. The monkeys are assumed to have only partial knowledge and only partial (or *bounded*) rationality. Consequently, the outcomes of possible interventions depend crucially on the *exact* assumptions made about the internal representations and processing of the agents (monkeys), particularly their representation of the room, their ability to create and execute plans of certain types and their ability to pass certain types of messages. There are no easy generalisations. However, it is clear that spontaneous co-operation is far from likely. Furthermore, misinformation should not be underestimated as an intervention strategy. However, intervention by lying is clearly less simple and less effective in real life than this little scenario suggests. There seems to be a need for a theory of *plausible* lying to constrain what we what we may take to be the set of effective interventions.

It should be noted in passing that there is a clear relationship between the Extended MandB scenario and the well-known *Prisoner's Dilemma* of game theory. If we take "co-operation" to mean that a monkey takes part in a joint plan to obtain bananas, and "defection" to mean that it does not do so, then the payoffs correspond to the standard PD pattern. Defection is always the best thing for a monkey to do, assuming that the payoffs of mutual defection are averaged over successive games. But representing the scenario as a single PD does not begin to do justice to it. For example, a monkey may co-operate to obtain the bananas and then defect when it comes to sharing them. Much more important is the issue of partial knowledge and partial rationality and its impact, and the connection to misrepresentation and mis-belief (Doran 1998).

8. A Macro-Model Of Watershed Intervention Strategies

The extended MandB scenario was introduced as a *micro-model* for the obvious reason that, although it addressed certain notions of co-operation, sustainability and intervention, it did so for a system of only two agents. By contrast we now sketch a macro-level model for watersheds, such as that of the Fraser River, and of the *intervention strategies* by which a watershed ecosystem (including its human components) may be acted upon. Some preliminary remarks are needed. Firstly, *the model is developed subject to the conceptual constraint that it must be programmable* for a computer. Secondly, the previous remark notwithstanding, *we make no attempt to state the model in any rigorously formal way*. To do so at this stage would merely deflect us from essentials. Thirdly, *the model is highly abstract*. The aim is to map out how some degree of understanding of this type of problem may be obtained, rather than draw specific conclusions about the Fraser itself. Finally, the model is *people centred*, with people represented by software agents.

The model takes as its start point the assumption that there exist centralised hierarchies of agents, with overlapping spheres of action, which compete in an initially resource rich environment.

Major components of the model.

> There is a large society of software agents (>100,000). The agents are heterogeneous in the sense that different agents have different behaviours and beliefs, and different types and degrees of cognitive ability.
>
> The agents are organised into competing hierarchies. At the top of each hierarchy is an agent whose goals reflect the resource interests of the hierarchy as a whole. Agents elsewhere in the hierarchy normally act loosely in accordance with more limited goals set by their immediate superiors.
>
> The society is set in an environment that incorporates an abstract resource management task (the "watershed resource model") that impacts all agents. Solutions may be assumed to require extensive agent co-operation. The hierarchies have overlapping zones of activity.
>
> Intervention is possible in the sense that the "experimenter" may send messages to some or all of the agents within the society.

To take this specification towards an actual computer implementation, we comment briefly on each of these four components in turn.

The agents. As was indicated earlier, there exists a range of available agent designs which might be adopted for this model but none entirely without problem. Agents should include beliefs, goals, and to some degree internal models of their "social" context and the environment. They need the ability to act, perhaps via the creation and execution of plans, and the ability to send and receive messages. One way to express these requirements is to say that agents must have an operational *Belief, Desires and Intentions (BDI)* architecture, including the ability to process incoming and outgoing speech-acts. Such agent architectures have been developed but still with cognitive abilities very limited compared with human beings.

The agent hierarchies. There are several ways in which a set of software agents may be engineered to function as an organised hierarchy. A simple method is to include within each agent specific behavioural rules that straightforwardly lead it to play its required role within the organisation. This role will include responding appropriately to messages from subordinates, and the adoption and pursuit of goals passed to it from superiors. The limitation of this method is that the agents are not themselves able to vary, or withdraw from, their own role, so that the hierarchy is inflexible. A different and more complex approach is to design agents in terms of the "contracts" that they negotiate, assess, agree, and adhere to. An agent's role in an organisation is then the consequence of the set of long-term contracts that it has accepted that lead it to behave in accordance with organisational goals and procedures. However, an agent has to be "persuaded" to undertake, and to remain within, an organisational role, essentially by reference to its own self-interest. Thus this approach is more sophisticated, and the hierarchies formed more flexible, but is also much more complex because it requires that agents be designed and implemented to manage contracts as a primary cognitive activity.

The watershed resource sub-model. The watershed resource model must address such key issues as pollution control and the need for "sustainability", and hence capture the need for both *local and global action* by agents. It should also permit the issue of equality or inequality of agent access to resources to be examined. A suitable form for the watershed resource model might be set of variables as follows:

- environmental variables (which are stochastically determined e.g. an input flow)
- sensed variables (variables one of more agents can read, to be interpreted as e.g. pollution level at a particular point)
- resource consumption variables (variables one or more agents can diminish e.g. consumption of water)
- input variables (variables one or more agents can augment e.g. amount of pesticide)
- action variables (actions one or more agents can perform e.g. specifying the flow through a dam)

together with a physical model (for example, a set of non-linear recurrence relations) which specifies the interconnections between the variables. Notice that spatial distribution would be implicit in the model structure. The "sphere of influence" of an agent (or agent hierarchy) would then be that subset of all the variables accessible to it.

The intervention language. The intervention language must specify what types of messages the experimenter may send to agents. It could be a widely used inter-agent communication language such as KQML (Weiss 1999), or a more restricted set of messages perhaps limited to assertions and questions. Whatever intervention language is used, agents must be designed to receive, assimilate and act upon -- or decline to act upon -- the messages and their content. This may well involve complex issues of belief and goal adjustment.

8.1. POSSIBLE INTERVENTION STRATEGIES

What can be said about strategies for intervention in the watershed model in pursuit of sustainability and equity -- keeping in mind our earlier investigation of the extended MandB scenario (section 7)? Recalling that we are considering systems of *software* agents, the following classes of intervention strategy easily come to mind:

- Take direct control of all agents, or perhaps just of "heads of hierarchies", giving them detailed and suitable instructions. This implies sending *goal messages* to the agents at issue, and the surely questionable assumption that these goals will be accepted by agents that by definition have been designed with a degree of autonomy.
- Perhaps after questioning, send *assertion messages* to the agents to augment their knowledge, with the intention that the agents, now more knowledgeable, work out an effective course of action for themselves.
- Follow an ideological/misinformation strategy, that is, send *assertion messages* to the agents which provide them with plausible but partially inaccurate or false information, the cumulative effect of which is nevertheless to bring about the desired collective action.
- Send messages the effect of which is to de-stabilise the agent system, intending or hoping that it will later self-organise more effectively.

Strategy (1) looks implausible, even for software agents. Strategy (2) assumes that fully knowledgeable agents will act in their long-term collective interest, and also fairly, which is also a questionable assumption. Strategies (3) and (4) are more complex and "tricky", but are nonetheless familiar in human terms.

The central challenge is to spell out these or similar strategies in detail, and to experiment with them to assess their real effectiveness. This is clearly a major task. Even if strategies can be precisely formulated in terms of the specified intervention language, their effectiveness will likely depend greatly upon details of the agents (as in the extended MandB scenario), the agent hierarchies, and the resource problem.

It is important to appreciate that here is an opportunity to assess possible intervention strategies, and perhaps to generate new and effective ones, from the model alone. This could be done, at least in principle, by automatically searching the space of all possible intervention strategies, perhaps using some kind of genetic algorithm. The performance of a strategy would be assessed by the effectiveness of the intervention achieved as measured in the watershed model. Ultimately the aim would be to discover new and useful concepts in terms of which to discuss and formulate intervention strategies.

8.2. CENTRALISATION AND DECENTRALISATION

It seems natural for the choice of an intervention strategy to be preceded by the choice of an organisation for the agents in the system -- effectively the level II formulation discussed in section 6.1. Therefore as a step towards consideration of effective intervention strategies, we briefly comment on centralisation and decentralisation and criteria for their effectiveness, making some observations about the impact of hierarchies on collective task performance (compare Prietula, Carley, and Gasser 1998).

From a computational perspective, centralised hierarchical organisation potentially enables one agent to possess and take decisions in the light of a single abstract overview of the problem scenario, and this may be of crucial benefit to efficient collective task performance. But this benefit will not arise if

- the task simply does not require such an overview for its performance (for example, if local problems do not inter-relate), or if
- no adequate abstraction scheme exists (implying consequent information overload at the centre if an attempt is made to form an overview), or if
- the flow of information up the hierarchy and of action down the hierarchy frequently fails or is too slow.

The purpose of *decentralisation* should thus be to recognise and address these difficulties with centralisation by striking a balance between the actual need for and feasibility of central overviews in any particular case, and the delays and errors inherent in any substantial attempt to create them. From this perspective intervention in the model needs to (a) establish what level degree of centralisation is the most effective and then (b) aim to achieve it. Some combination of enriching the lower level knowledge content and connectivity of individual hierarchies, enhancing inter-hierarchy information flow, and filtering or merging the existing set of hierarchies might do the latter.

8.3. TESTING THE WATERSHED MODEL

It is a *potential test of the proposed watershed model* to ask if effective intervention strategies derived from it (see section 8.1) can be related to those in current practice and in the relevant literature in an insightful way. If so, the model begins to look like a useful tool.

In fact, a number of ecosystem management strategies are prominent in the environmental resource management literature reflecting the current emphasis on decentralised and co-operative ecosystem management. These include:

- Maximal Stakeholder involvement (e.g. Dorcey 1997)
- Co-management (e.g. Nielsen and Vedsmand 1999)
- Community Management (e.g. Ostrom 1990, Christie and White 1997, Lansing, Kremer and Smuts 1998, Lansing 2000)
- Adaptive Management (Walters 1986, Healey 1998)

These different strategies may be summarised as follows:

> *Maximal Stakeholder Involvement* envisages a situation in which there are multiple competing stakeholders for a particular ecosystem (e.g. various branches of government, non-governmental organisations, and corporations). These must be drawn into discussion, negotiation and co-operation partly to achieve better decision making by way of better mutual understanding and shared knowledge, and partly to achieve greater commitment to implementation of solutions.
>
> *Co-Management* promotes co-operation between government agencies and local communities, for example for fisheries management.

Particularly in the context of non-industrialised countries:

> *Community Management* seeks to draw upon (or return to) local community structure (a form of *social capital*) as a basis for resource system management in contrast to the involvement of government, with emphasis upon the effectiveness of local knowledge and social regulation.

By contrast:

> *Adaptive Management* is more concerned with the "how" of management than the "who". The emphasis is on a (new) management style in which experimentation is at the heart of management practice so that information is gathered about the ecosystem as management proceeds, and action is varied accordingly.

Although not the focus of our attention here, a further possibility is merely to act to *improve the resource sub-system*, without seeking to change, or even understand, the associated human socio-cultural system. There is much evidence to suggest that this approach can prove counter-productive (Ostrom 1995).

These various management strategies, although all in practical use, are not all of the same type, are not precisely defined, and are not exhaustive or well differentiated. Their limits of applicability are little considered in the literature except in the vaguest terms. The question naturally arises, therefore, which of these strategies should be used in which circumstances? And what other possible ecosystem management strategies are there? It is precisely such questions that the macro-model sketched here might go some way towards answering.

In the light of the analysis to this point we may offer the following more specific remarks. Traditional centralised management strategies, together with the *Maximal Stakeholder Involvement* and *Co-Management* strategies just described, are surely best analysed in close relation to the issues around agent hierarchies considered in section 8.2. Furthermore, it may be assumed that all involve elements of knowledge provision, implicitly or explicitly. *Adaptive Management* seems essential if there is either initial ignorance of the details of the management task, or if the task is changing. Why might adaptive management *not* be employed? The answer surely lies in the notion of *cognitive economy*. To minimise computational outlay, it is always attractive for an agent (software, human or organisational) to commit to a course of action and not reconsider it other than in exceptional circumstances. Of course, this strategy of commitment can go seriously wrong if the reluctance to adapt to changing circumstances is inappropriate or excessive. The *Community Management* strategy is somewhat different. It puts the emphasis heavily upon distributed management and organisation, and rather little on hierarchies, and therefore involves multi-agent system co-operation regimes that do not construct any overview. Finally, the models considered here suggest that the effectiveness of *Machiavellian* or *Trojan Horse* strategies should not be underestimated.

9. Discussion: The Methodological Problems Revisited

The approach we have followed to address the core problems of Fraser River watershed management has been determined by the observation that the heart of the problem is to change the behaviour of people. This has led us away from low-level models (i.e. level I and level II models) towards the modelling of strategies of intervention and their effects. The outcome is an outline agent-based model of the Fraser River intervention scenario, which in full specification would be very abstract as compared with the target system. It would contain many essentially arbitrary structural details and adjustable parameters both within and external to agents. All the methodological problems discussed earlier in this paper therefore arise. Why use an agent-based model at all? Why work at this level of abstraction? How can the model be validated? How can the regularities in its behaviour be established, given its massive parameter space?

We argue that use of agent-based models follows from the need to address the essential processes of human social systems if the important environmental management questions are to be answered. The complexity of the model is inevitable given the nature of the target system and the need to address what agents believe as well as what they do. And the model's complexity in turn implies that component by component validation is impossible and macro-validation can give us no confidence in the ability of the model to predict in new circumstances -- rather we *know* that it is incorrect. However, by searching the parameter space, it may well be possible to discover macro-regularities of significance that may or may not be expected.

10. Conclusions And Prospects

As the various models discussed in this paper illustrate, agent-based modelling in the social and environmental sciences seems to have much to offer. Given the importance, in particular, of the environmental problems to be tackled this potential must surely be followed up. But agent-based modelling is at present intensive in specialist knowledge and encounters major practical and methodological difficulties that will not be overcome without substantial technical research. It is no surprise, then, that there have been virtually no agent-based social modelling studies to date that have gone beyond "proof of concept", or "stimulation of discussion", or, at best, bringing forward a few interesting generalisations of uncertain validity.

So CAN agent-based modelling really be useful? The answer seems to me to be that the "soft" uses of agent based modelling, such as its use to stimulate and focus discussion amongst stakeholders from a new agent-based perspective, has unquestionable practical value now. Furthermore, I believe that much progress could be made now were *adequate and sufficiently expert resources deployed* on projects such as the model of watershed ecosystem management proposed here. There is a good chance that a major, systematic and detailed investigation along the lines indicated would yield real insights of practical value. The difficulty is that the human resources required for such a study run into many skilled man-years, especially of expertise in the design and implementation of "intelligent" agents in a social simulation context. This is not easily come by when the relevant human expertise is much in demand for more straightforward and immediately profitable software engineering projects.

Finally, it does seem that our current inability to program cognitive processes to anything approaching human level, still less anything that looks like "general intelligence" could be a major obstacle in an agent based study of ecosystem intervention strategies. Persuasion, of its essence, seems to relate to *all* aspects of cognition. In this connection, we should perhaps note and question the habitual assumption that software agents should be designed to behave rationally (if self-interestedly), even when they are created in support of social modelling. Are less-rational aspects of human behaviour really of no modelling significance? The impact both of *habitual* patterns of behaviour and thought and of the *social emotions*, surely merits careful study and may well provide crucial extensions to the agent-based models currently being explored. Leadership, in particular, is not just about rational decision-making, but also requires understanding of the needs and emotions of followers. Recently some small steps have been taken in this direction (Doran 2000; Norling, Sonenberg and Ronnquist, 2001).

Acknowledgements

I gratefully acknowledge hospitality provided by the Institute for Resources and the Environment of the University of British Columbia, Canada in February and March 2000, which gave me the opportunity to study in some depth the Fraser River watershed and its associated management problems.

References

Axtell, R. (2001). Effects of Interaction Topology and Activation Regime in Several Multi-Agent Systems. In: 33-48.

Bousquet, F., Cambier, C., Mullon, C., Morand, P., and Quensiere, J. (1994). Simulating Fishermen's Society. In: Gilbert, N. and Doran, J.E. (eds.) *Simulating Societies*. UCL Press, London. pp. 143-164.

Carpenter, S., Brock, W. and Hanson, P. (1999). Ecological and Social Dynamics in Simple Models of Ecosystem management. *Conservation Ecology 3(2): 4.* http://www.consecol.org/vol3/iss2/art4

Christie, P. and White, A.T. (1997). Trends in Development of Coastal Area Management in Tropical Countries: From Central to Community Orientation. *Coastal Management, 25,155-181.*

Doran, J.E. (1997). From Computer Simulation to Artificial Societies. *Transactions of the Society for Computer Simulation International*, 14, 69-78.

Doran, J.E. (1998). Simulating Collective Misbelief. *Journal of Artificial Societies and Social Simulation, Vol 1(1)* http://www.soc.surrey.ac.uk/JASSS/1/1/3.html

Doran, J.E. (2000). Trajectories to Complexity in Artificial Societies: Rationality, Belief and Emotions. In: T. A. Kohler and G. J. Gummerman (eds.). Dynamics in Human and Primate Societies. Oxford University Press, Oxford. pp. 89-105.

Doran, J.E. and Palmer, M. (1995). The EOS Project: Integrating Two Models of Palaeolithic Social Change. In: Gilbert, N. and Conte, R. (eds.). *Artificial Societies*. UCL Press, London. pp. 103-125.

Dorcey, A.H.J. (1997). Collaborating Towards Sustainability Together: The Fraser Basin Management Board and Program. In: Shrubsole, D. and Mitchell, B. (eds.). *Practising Sustainable Water Management: Canadian and International Experiences*. Canadian Water Resources Association. Available at http://www.interchg.ubc.ca/dorcey/chcwra/fccwra.html

Downing, T.E., Moss, S., and Pahl-Wostl, C. (2001). Understanding Climate Policy Using Participatory Agent-Based Social Simulation. In: Moss, S. and Davidsson, P. (eds.). *Multi-Agent Based Simulation*. Springer, Berlin. LNAI 1979. pp. 198-213.

Epstein, J.M. and Axtell, R. (1996). *Growing Artificial Societies: Social Science from the Bottom Up. Washington, D.C.* The Brookings Institution Press and MIT Press, Cambridge, MA.

FIRMA (2000). See web page at http://www.cpm.mmu.ac.uk/firma/index.html

Fishwick, P.A., Sanderson, J.G. and Wolff, W.F. (1998). A Multimodeling Basis for Across-Trophic-Level Ecosystem Modeling: The Florida Everglades Example. *SCS Transactions on Simulation*, 15(2), 76-98.

Gilbert, G.N. (2000). Modelling Sociality: The View from Europe. In: T. A. Kohler and G. J. Gummerman (eds.). *Dynamics in Human and Primate Societies*. Oxford University Press, Oxford. pp. 355-372.

Gilbert, G.N. and Troitzsch K.G., (1999). *Simulation for the Social Scientist*. Open University Press.

Hales, D. (2001). Memetic Evolution and Sub-Optimisation. PhD Thesis, Department of Computer Science, University of Essex, Colchester, UK

Hall, K. *et al* (2000) Brunette Basin Watershed Plan. Policy and Planning Department, Greater Vancouver Regional District.

Healey, M. (1998). Paradigms, Policies, and Prognostications about the Management of Watershed Ecosystems In: Naiman, R.J. and Bilby, R.E. (eds.) *River Ecology and Management*. Springer. Pp. 662-682.

Healey, M. (1998a). Barriers and Bridges to Sustainability in the Fraser Basin. Invited address to the *State of the Basin conference*, November 20 and 21st, 1998. Vancouver.

Healey, M. (ed.) (1999). *Seeking Sustainability in the Lower Fraser Basin: Issues and Choices.* Institute for Resources and the Environment, Westwater Research Centre, University of British Columbia, Vancouver.

Jennings, N.R., and Wooldridge, M.J. (eds.) (1998). *Agent Technology*. Springer, Berlin.

Kluver, J. and Schmidt, J. (1999). Topology, Metric and Dynamics of Social Systems. *Journal of Artificial Societies and Social Simulation*, vol. 2, no. 3, http://www.soc.surrey.ac.uk/JASSS/2/3/7.html

Kohler, T.A., Kresl, J., Van West, C., Carr E., and Wilshusen, R.H. (2000). Be There Then: a Modelling Approach to Settlement Determinants and Spatial Efficiency Among Late Ancestral Pueblo Populations of the Mesa Verde Region, U.S. In: T. A. Kohler and G. J. Gummerman (eds.). *Dynamics in Human and Primate Societies*. Oxford University Press. pp. 145-178.

Lansing, J.S. (2000). Anti-Chaos, Common Property, and the Emergence of Cooperation. In: T. A. Kohler and G. J. Gummerman (eds.). *Dynamics in Human and Primate Societies*. Oxford University Press. pp. 207-223.

Lansing, J.S, Kremer, J.N, and Smuts, B.B. (1998). System-Dependent Selection, Ecological Feedback and the Emergence of Functional Structure in Ecosystems. *J Theor. Biol.* 192, 377-391.

Lawson, B.G. and Park, S. (2000). Asynchronous Time Evolution in an Artificial Society Model. *[ONLINE] JASSS, Vol. 3, No. 1.*

Lee, K. and Fishwick, P.A. (1997). A Methodology for Dynamic Model Abstraction *Transactions of the Society for Computer Simulation International,* 13(4), 217-229.

Marshall, D. (1998) Watershed management in British Columbia: The Fraser Basin Experience. *Environments*, Vol. 25, No. 2/3, 64-79.

Moss, S. (1998). Critical Incident Management: An Empirically Derived Computational Model. *[ONLINE] JASSS, Vol. 1, No. 4.*

Moss S. (2001). Messy Systems: The Target for Multi Agent Based Simulation. In: Moss, S. and Davidsson, P. (eds.). Multi-Agent Based Simulation. Springer, Berlin. LNAI 1979. pp. 1-14.

Nielsen, J.R. and Vedsmand, T. (1999). User Participation and Institutional Change in Fisheries Management: a Viable Alternative to the Failures of 'Top-Down' Driven Control? *Ocean & Coastal Management*, 42, 19-37.

Nilsson, N.J. (1971). *Problem-Solving Methods in Artificial Intelligence*. McGraw-Hill.

Norling, E., Sonenberg, L. and Ronnquist, R. (2001). Enhancing Multi-Agent Based Simulation with Human-Like Decision Making Strategies. In: Moss, S. and Davidsson, P. (eds.). Multi-Agent Based Simulation. Springer, Berlin. LNAI 1979. pp. 214-228

Ostrom, E. (1990). *Governing the Commons: The Evolution of Institutions for Collective Action.* Cambridge University Press, Cambridge.

Ostrom, E. (1995). Constituting Social Capital and Collective Action. In: R. O. Kehane and E. Ostrom (eds.). *Local Commons and Global Interdependence: Heterogeneity and Cooperation in Two Domains*. Sage Publications, London. pp. 125-160.

Prietula, M.J., Carley, K.M., and Gasser, L. (1998). *Simulating Organisations*. AAAI and MIT Press.

Rouchier, J., Bousquet, F., Le Page, C., and Bonnefoy, J.-L. (2001). Multi-Agent Modelling and Renewable Resource Issues: the Relevance of Shared Representations for Interacting Agents. In: Moss, S. and Davidsson, P. (eds.). Multi-Agent Based Simulation. Springer, Berlin. LNAI 1979. pp. 181-197.

Russell, S. and Norvig, P. (1995). *Artificial Intelligence: a Modern Approach*. Prentice-Hall.

Teran, O., Edmunds, B., and Wallis, S. (2001). Mapping the Envelope of Social Simulation Trajectories. In: Moss, S. and Davidsson, P. (eds.). Multi-Agent Based Simulation. Springer, Berlin. LNAI 1979. pp. 229-243.

Walters, C.J. (1986). *Adaptive Management of Renewable Resources*. Macmillan, New York.

WCED (1987). *Our Common Future*. Oxford University Press, Oxford.

Weiss, G. (ed.) (1999). *Multiagent Systems*. The MIT Press, Cambridge, MA and London.

FRAUD, EXCHANGE AND CO-OPERATION

RITA CECCHI
Department of Communication Sciences
University of Siena
Siena, Italy

and

ROSARIA CONTE and CRISTIANO CASTELFRANCHI
PSS (Project on Social Simulation)
Division for "AI, Cognitive and Interaction Modelling"
National Research Council, Institute of Psychology
Rome, Italy

Abstract

This paper will provide an analysis of the cognitive plans that lead to pro-social or individualistic action. It will demonstrate that the process of exchange has a natural tendency to fraud regardless of external controls and detection, while co-operation is a fundamentally pro-social action. The paper will also demonstrate that external observation cannot determine whether a physical exchange of goods can be considered as an exchange in the cognitive sense: there is no need for an exchange of material goods to create this structure in the agents' mind. By contrast, co-operation is defined by a shared goal and can be present in the agent's mind even if there is no external action to validate it. A cognitive approach to co-operation and exchange allows us to show precisely why (and under what circumstances) the structures of these relationships are differently resistant to fraud and deception.

1. Introduction

In social theory, especially in formal models of social action (game theory), the social world is essentially viewed as consistent with the *bellum omnium contra omnes* principle. In other words, the social world is characterised as:

- a domain of interference and collisions
- which calls for coordination among the agents

Indeed, co-operation is usually explained by the prospect of re-encounter. To use the words of Axelrod (1997, 12), game theoretic models explain co-operation in the *shadow of the future*, as a forward-looking, calculated and self-

N.J. Saam and B. Schmidt (eds.)
Cooperative Agents. Applications in the Social Sciences, 83-107

interested[1] decision to co-operate. (See also Macy 1998.) If the one shot Prisoner's Dilemma (PD) game leaves no room for co-operation, repeated versions of the same game do (see Axelrod 1984).

In this paper we will endeavour to do justice to the intuition that social action also varies with the probability of defection. More precisely, while exchange is always jeopardised by the probability of deception and fraud, co-operative action is relatively immune to the risk of anti-social behaviour. In contrast to the view supported by economic rationality and game theory, social life offers considerable opportunities for co-operation and even for one shot co-operation. By contrast, exchange is intimately intertwined with deception and fraud, to the point that one is impossible without the other. To see why, it is necessary to ground models of exchange and co-operation in a general theory of action. This theory necessitates a cognitive (or Artificial Intelligence) framework. (For the most classical version of a theory of planned action, see Miller et al. 1960.) The following theses will be discussed in the paper:

- There are different types of positive or pro-social action which include co-operation and exchange.
- A plan based model of action is necessary to *disentangle co-operation from exchange*. While the PD game applies to the latter, it *never* applies to the former (Conte et al. 1999). Even in its repeated version, the PD cannot account for co-operation as distinct from honest bargaining.
- The plan based model is also necessary to account for the different role of defection in co-operation and exchange.

A subsidiary question is why we need a more analytical view of social action. In fact, it does not really matter whether it is exchange or co-operation that occurs. The point is whether or not equitable social action, satisfactory for all the parties involved, can occur spontaneously. If it can, under what conditions will it occur? Our answer is that it can occur spontaneously under given conditions. However, social action is not only conditioned to the probability of re-encounter, as game theorists claim, but also to the structural social relationships between the agents' goals and abilities. Society is not a necessary evil, which limits the autonomy, powers and achievements of individual agents and therefore calls for deception, fraud and cheating. Unfortunately, this view is deeply rooted in the Western conception of mentality and in the utilitarian tradition.[2] In practice, society is also a *resource* for agents that are limited on their own and cannot satisfy some of their goals

---

[1] Although this self-interest is "enlightened", a point clarified by Binmore 1994.

[2] Consider the scientific impact of the von Neumann and Morgenstern theory of decision. This approach was inherited by Social Artificial Intelligence (Distributed Artificial Intelligence and Multi-agent systems). It is now influential in social simulation too. (See Axelrod 1984, Axelrod and Dion 1988, Hirshleifer and Coll 1988, Molander 1985, Mueller 1987, Rubinstein 1986 and Bicchieri 1990.) It should be noted, however, that the overall results obtained from these studies have been shown (cf. Lomborg, 1994) to be neither consistent nor satisfactory.

without others' help. Depending on these structural relationships, agents can sometimes cheat benefactors (namely, when agents need others to achieve their own private goals), but at other times cannot (when agents need each other's help to achieve one and the same goal). Therefore, a theory of structural social relationships can account for the necessity and occurrence of deception and fraud in social life.

In the final section (§ 5) of the paper, we will return to a discussion of the type of analysis presented here. Before that, we will discuss the rational theory approach to the subject matter (§ 2), and present our models of co-operation (§ 3.1) and exchange (§ 3.2). Then, we will present a general notion of deception and fraud (§ 4.1) and apply it to both types of social action. The analysis presented will show why (and when) deception and fraud are necessary in exchange but not in co-operation.

2. A Rational Approach to Pro-social Action

Game theory sees the social world as a "Hobbesian game", a generalised PD in which agents constantly interfere with one another and therefore need to resort to agreements to prevent mutual disaster. In the game theoretic view, society is a realm of interference (positive or negative) among agents, a situation referred to as *strategic interdependence*. According to such a view, any agent's choice bears consequences for all others. To paraphrase Axelrod, given a multi-agent context, there is no *a priori* best strategy because the outcome of each strategy necessarily depends on the strategies of the other players. According to this conception, social relations are only the effects of individual choices. They are not seen as conditions for it. This is essentially due to the coarse grained specification of agents which game theory implicitly provides. In particular:

- Agent's goals are ignored. In AI, the agent is a planning system and goals are essential to explain why and how actions are planned. In game theory, by contrast, actions are explained *directly* in terms of the rational choice principle (Rasmusen 1990, Binmore 1987, Luce and Raiffa 1989). Given perfectly rational agents, game theory can predict the point where no agent unilaterally wishes to change their choice (Nash Equilibrium). The rational choice principle is a fundamental mechanism that characterises self interested deciding systems. However, it is not the primary motor for deciding systems. Agents are not motivated to action by the principle of maximum utility even though their actions may be under control of this principle. Actions are ruled by needs and desires and controlled by goals. Analogously, in social situations, people do not face the option to co-operate or defect. When they decide to co-operate, they do so to achieve some pre-existing goal. Of course, if co-operation is expected to be more costly than beneficial, it will be discarded. However, the question is why co-operation is considered at all, rather than why it is chosen or rejected. Game theorists do not address this question. Since goals are ignored, the agents' reasons for actively engaging in social actions (which are the preconditions for social choice) are ignored as well.

- Agents are seen as practically homogeneous except that they may resort to different interaction strategies. (See Axelrod 1984 for an overview, Rasmusen 1990, Bicchieri 1990 and Lomborg 1994). However, agents are not seen as needing one another's help. Instead, they are seen as perfectly at ease on their own. Society seems to put together a bunch of self-sufficient beings that merely ought to take one another's doings into account. Agents are essentially well adjusted and powerful. It is the social world that limits their powers and reduces their degree of adaptation.

In the following discussion, we will argue that a number of problems and inadequacies (more or less acknowledged by the game theoretic approach) are caused by this view of sociality. The inadequacies are the following:

- **The dilemma of cooperation**: game theoretic studies on the emergence of cooperation have not produced consistent and satisfactory results (Lomborg 1994). The impossibility of justifying co-operation in one shot PD games is largely recognised. Some authors (Howard 1971, Gauthier 1986, 1993) have attempted to extend the PD to account for one shot co-operation but others (Binmore 1994) argue strongly that such extensions involve irrationality or faulty logic. Is it ultimately possible to account for cooperation in a game-theoretic framework?
- **Behaviours between co-operation and defection**: social choice is richer and more differentiated than game theory allows.
- **Strategic beliefs versus social goals**: for game theorists, social agency is equivalent to strategic agency. The latter is defined as taking the effects brought about by other agents into account. But social agents do something more than simply taking others into account. They also act upon others. How can this fact be accounted for in a game theoretic framework?

In the subsequent sections, we will examine each of these problems in turn.

2.1. THE DILEMMA OF CO-OPERATION

What game theorists do not seem to have explained is how co-operation is possible. The well-known argument is as follows. In repeated games, the higher the number of players - and consequently the lower the probabilities for the same agents to interact repeatedly - the more destructive (either jointly or separately) are the players' options. Consider a classical payoff matrix for the PD game.

		Player 2	
		C	D
Player 1	C	1 1	-2 2
	D	2 -2	-1 -1

Table 1: The Prisoner's Dilemma from Ullman-Margalit (1977, 18)

Here, the ideal situation for each player is to play Defect when the opponent has played Co-operate. The obvious question is how a co-operative answer is possible in these circumstances. Game theorists have generally searched for a strategic solution to this question involving repeated play and/or absence of mutual knowledge among the players. In other words, a co-operative choice can be expected in two situations:

- If players want to prevent their opponents' defection by showing a co-operative "attitude"
- If players are uncertain as to the rules applied by their opponents. For example, player 1 might be expected to play Co-operate if he did not expect player 2 to follow the rational rule indicating defection as a response.

Conversely, a co-operative answer is less likely to occur under the following conditions:

- Low chances of punishment: the expected probability that one's defection will be "punished" by the other players is low
- Free-riding: the expected probabilities of participating in a joint positive outcome are relatively independent of the choice one makes. This happens in large groups and also in the provision of public goods. Even if you do not pay the municipal fee for street lighting, you can expect to enjoy it all the same.

Moreover, it can easily be shown that the non co-operative outcome tends to become increasingly stable. The more the agents play Defect, the more rational it is for them to persist in this option and for other agents to join them.

In a game theoretic account, agents' choices tend to produce a state of affairs that is both individually and socially undesirable.

The game theoretic explanation of co-operation is not much different from the game theoretic view of norms and morals (cf. Axelrod 1986). Indeed, game theory has trouble explaining both phenomena (Heckathorn 1988). Only by integrating an evolutionary dimension (Maynard Smith 1982) can stable co-operative effects be observed. These models have been applied to the emergence of co-operation (Lomborg 1994) and norms (Bicchieri 1990).

To sum up, game-theory regards co-operation is a useful social device, a solution to a social conflict or problem. But there is another sense of co-operation that game theorists do not and cannot account for. Co-operation can also be a means for achieving one's goals, for solving individual problems.

2.2. BETWEEN CO-OPERATION AND DEFECTION

Essentially, game theorists conceive of social interaction in terms of interference thus calling for conventions and norms of coordination. This conception does not allow several types of social actions to be modelled: goal adoption, co-operation, social exchange and so on. It does not account for the diverse reasons why self-interested agents engage in these kinds of social action. A view of social action grounded in interference is not able to differentiate these types of behaviour and reasons for engaging in them. An analysis of the different types of social action also has practical advantages at least from an AI planning viewpoint. In order to design autonomous artificial agents which actively plan (social) action, it is necessary to develop a theory of agents' motives for participating in social interaction (cf. Conte 1999).

The different forms of positive interaction represent different social solutions to problems that individuals happen to face. Let us distinguish at least two of them:

- Exchange: this occurs when agents comply with others' requests in order to obtain reciprocal compliance
- Co-operation: agents each agree to accomplish a share of a multi-agent plan that is designed to achieve a shared goal (Conte and Castelfranchi 1995).

Only the former action fits the structure of the Iterated Prisoner's Dilemma (IPD) where agents are expected to co-operate in order to avoid retaliation. However, the notion of exchange that we have in mind does not imply any social conflict or interference. According to our conception, social exchange is a means for each agent to solve his or her own problem. Social exchange is therefore a sub goal in a plan for an individual goal. It represents a rational solution to an individual (as opposed to social) problem. The solution is social, but the problem is not, since it does not derive from interference and conflicts among agents.

Co-operation, at least in an AI planning perspective, implies a multi-agent plan for achieving a common goal. The difference between co-operation and exchange is not merely theoretical.

- It is reflected in the agents' plans: while exchange entails no common plan but only involves two intertwined plans, co-operation necessarily entails a common multi-agent plan.
- It implies a different tolerance of cheating: in exchange agents do not truly share interests, while in co-operation they do. Hence, while cheating can be a rational choice in exchange (as is well-known in game theory), it is essentially irrational in co-operation since it would lead the cheat to harm the execution of his or her own plan.

2.3. STRATEGIC BELIEFS VERSUS SOCIAL GOALS

The most fundamental difference between the game theoretic and planning perspectives on social agency probably concerns the notions of "social", "social agent", "social choice" and so on. In game theory:

- Social means strategic: a social agent is a strategic agent and a social choice is a strategic choice, taking into account the actions and choices of the other players.
- The term "social" is assigned to beliefs: a social action or choice is guided by beliefs about the moves of others.

This is an *insufficient* way of characterising sociality. To cite Piaget, it emphasises the role of *accommodation* but underestimates the function of *assimilation*. It is true that one takes others into account if one is to accommodate personal behaviour to the social environment. But a social agent can do something more. It can try to modify others' behaviour, by influencing their mental states (assimilation). Indeed, retaliation is a means for modifying the strategies of the other players. But what game theorists call retaliation is a strategically rational choice rather than a truly social action. It is a choice which takes others' choices into account rather than attempting to modify the other agents' behaviours by influencing their minds.

In fact, agents do not limit themselves to predicting what their opponents will do. They also act upon one another in many direct and indirect ways. This has an enormous impact on social life and may have interesting applications in the construction of artificial systems for distributed planning, problem solving and so on.

3. A Cognitive Approach to Pro-social Action.

A cognitive agent is relatively autonomous from external stimuli. Its beliefs and goals mediate its interaction with the world. It does not apply reactive routines. Instead, its actions result from a mental elaboration of external inputs.

A cognitive agent is autonomous when it has its own goals, according to which it accepts or rejects others' requests (see Conte and Castelfranchi 1995). A cognitive agent accepts others' requests if this is a means by which it can reach one of its own higher level goals.

In consequence, pro-social actions are not necessarily altruistic. Agent *x* can **instrumentally adopt** a goal held by agent *y* in order to obtain some favour in return. This is the case in an exchange relationship. But *x* can also **co-operate with** *y* to reach a shared goal (Conte and Castelfranchi 1995).[3] Plans for co-operation and exchange differ greatly and these differences in structure define the boundary between the two concepts.

A cognitive approach to pro-social action is fundamental to:

- Identify which social relationship we are considering. If we don't know the beliefs and goals in the agents' minds, we cannot tell whether it is co-operation or exchange that is occurring
- Analyse the influence that different relationships have on the agent's behaviours: their intention to deceive, their care in performing their actions and so on

Empirically, exchange and co-operation can converge. Sometimes two agents will co-operate in the interests of exchange or exchange with a view to co-operation. In such cases external observation does not help us very much.

3.1. A COGNITIVE APPROACH TO CO-OPERATION.

Co-operation is defined by a shared goal that can only be reached by mutual help between agents. In other words, co-operation involves a multi-agent plan, a single plan performed by two or more agents.

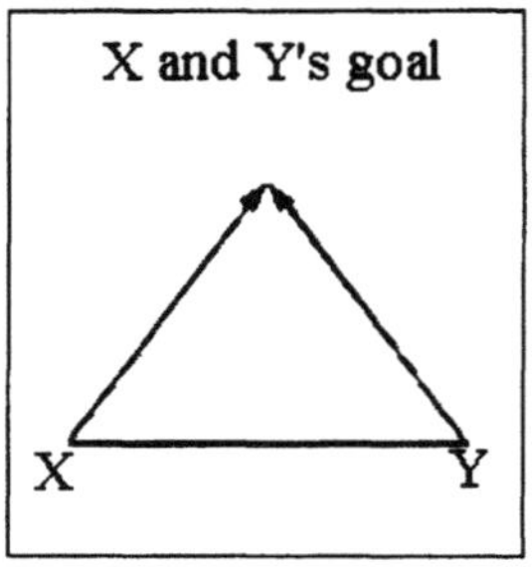

Figure 1

As we can see in *figure 1*, co-operation can be represented as a single plan. Agents are pursuing the same goal and, consequently, they are genuinely

[3] A third kind of behaviour is **terminal adoption**: *x* adopts a goal held by *y* without calculating any advantage or gain.

interested in the goal fulfilment of the other agent. If one agent does not reach its goal, neither will succeed. Moreover, each agent has a specific role within the plan and believes that the shared goal will only be reached by performing all the appropriate actions. **Agents mutually believe themselves to be mutually dependent.**

3.2. A COGNITIVE APPROACH TO EXCHANGE

By contrast, exchange is a means of reaching an individual goal and having one's own goal adopted by another.

In the social sciences, each discipline has used the notion of exchange according to its own definition. However, all of these definitions ignore the unobservable aspects of the exchange process: what goes on in the agents' minds. Our definition of exchange contains two main elements:

- Giving help to another agent, in order to
- Receive help

Agent *x* adopts the goal (P) of agent *y* with the expectation (and only on condition) that agent *y* will adopt the goal of agent *x* (S), and vice versa (*figure* 2).

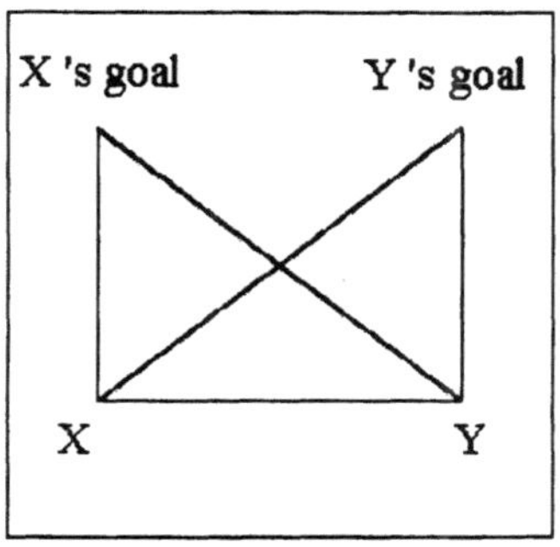

Figure 2

A situation in which one agent gives a good or resource to another (and receives something in return) is not necessarily an exchange relationship. The cognitive definition identifies the boundary between exchange and similar activities (instrumental adoption, social exchange, gift giving) which contain some of its elements but do not overlap completely. A simple external observation cannot help distinguish one concept from the other (Conte and Castelfranchi 1995).

Instrumental adoption is an attempt by agent *x* to influence agent *y* by adopting one of its goals in order to receive a particular favour. This relationship lacks mutually agreed reciprocal adoption that is essential to exchange. An expectation of exchange may exist in the mind of *x* but no duty on *y* corresponds to it. To see why, consider an invitation for dinner. The host can *hope* to be invited back but cannot enforce reciprocity. In fact, a loose link can exist between adoption and reciprocation: *x* adopts a goal belonging to *y* hoping to produce the adoption of one of its goals by *y*. However, there is neither necessary mutual obligation nor real reciprocation. It is a simple two-way goal adoption of the type often found in gift giving rituals. (The potlàtch of the Kwakiutl and the kula of the Trobriand islanders provide anthropological examples.[4])

When agents are involved in other kinds of relationships, a fundamental element of exchange is not represented in their minds, that of agreement. This agreement – "I will help you if you help me in return" - can be an implicit or explicit pact or a legally binding contract.

4. Deception: Option or Necessity?

According to classical economics, a sufficiently intelligent and rational agent has no reason to cheat.[5] In general, the theory of rational choice tends to eliminate deception from the range of possible solutions which an agent can find to a given problem. (See Muthoo 1999, Binmore and Vulkan 1997 and Binmore and Dasgupta 1987.)

In our opinion, this is not a valid perspective for two main reasons. Firstly, deceptive behaviour can be performed by a very complex mind. (In fact, in order to be successful, it often is!) It is hardly a sign of irrationality. Secondly, deception is often rational. Sometimes, it is the only means an agent has to reach a very important goal. A self interested cognitive agent, involved in a situation of exchange or in any other social activity, may have an objective incentive to cheat. This results from the particular structural conditions of the situation and from the expected costs and benefits involved. In weighing up pros and cons, the agent works out the possibility of future interactions with the same partner, the importance of reputation, the nature of punishment and the probabilities of detection and avoiding bad consequences.

Sometimes honest behaviour is worthwhile. The most significant example of this is co-operation to accomplish a goal which several agents have in common. The agents are each motivated to reach the goal of the other agents because that goal is shared. Sometimes, honesty is expensive but, considering all the goals and beliefs held by the agent (including ethical and normative goals), is more

---

[4] These are collective exchanges of gifts. They are useful (but not explicitly intended) ways of distributing resources within the population and creating social relationships that can be used in marriage, economic transactions, mutual aid and so on.

[5] "With people who are sufficiently intelligent and rational there should not be any question of 'bargaining ability', a term which suggests something like skill in duping the other fellow" (Nash 1953 cited in Binmore and Dasgupta 1987).

convenient than cheating. (This can be considered a sub case of the previous situation.) However, if the agent finds it vital to achieve the goal and/or cannot renounce it, they will engage in deception. This situation is typical of exchange relationships and is primarily caused by the scarcity of resources. In order to obtain a fundamental means, *x* can be compelled to deceive an exchange partner *y*, promising a resource *y* does not possess. This is a perfectly rational behaviour (Conte and Castelfranchi 1995).

Our claim is that exchange leads to deception more than other pro-social relations do. Furthermore, in some cases, it can be considered the only feasible solution.

4.1. DEFINITIONS OF DECEPTION AND FRAUD

In (Castelfranchi and Poggi 1998) deception is defined as an **act or feature** of an organism *L* (the liar) which has two properties. Firstly, the **aim** of withholding a **relevant true belief** from another organism *D* (the deceived). Secondly, **not to reveal the intention to deceive**.

This definition (broad enough to include deception among animals) needs further specification.

Acts and features share a teleological nature. They must be goal oriented. Deception then includes:

- Linguistic acts (lies)
- Non verbal communicative actions (a false smile)
- Non communicative actions
- Omissions
- Deceptive features: the morphological characteristics of some organisms are not intended by the organisms themselves to be deceptive. Nonetheless, they are oriented to an evolutionary end. The camouflage of the chameleon is functional to defend the animal from its predators. Furthermore, the notion we present encompasses objects used in (and produced for) deception. Examples include counterfeit coins and false beards.

As we have said, the ***teleological aspect of deception*** concerns both goals internal to the agent's mind and biological functions external to the agents. Here, we are interested in cognitive deception: a representation of the goal of deception in the mind of the agent.

Another concept that requires a specific explanation is ***true belief***. A cognitive agent is defined by its capacity to respond to external stimuli according to its beliefs and goals. For such an agent, a gap may exist between what is believed and "objective" truth. Therefore, if a cognitive agent is lying, it is giving false information about the content of its mind but not necessarily about the world. In this respect, we can define an agent as sincere when it does not conceal true beliefs and does not give false information regarding the content of its mind. When we are talking about non-cognitive agents, however, there is no difference between objective truth and subjective belief.

The ***relevance*** of concealed information is a fundamental element of the definition of deceit. It derives from the importance of the denied belief within the deceived agent's plan and is one of the most important ways to distinguish deception from a mere secret. In deception, *L* is committing an aggressive act towards *D* in two senses. Firstly, *L* does not let *D* reach its goal. Secondly, *D* is transgressing the general social law that prescribes knowledge sharing among agents.

Not making one's aim explicit is, ultimately, an element of deceit that cannot be eliminated. There are two levels of lying. At the first, we communicate false information or omit what is true. At the second, we hide our intention to deceive. This characteristic differentiates deception from irony. Irony makes provision for some communicative action denying the first level assertion. (This action is usually non-linguistic like a wry smile for example.)

The definition we have just provided has to be further specified in order to describe the forms which deception can take. For example, a lie is a linguistic form of deception while the act of hiding is a non-communicative one even if they are both goals represented in the agent's mind. The hierarchical structure of sub categories in deception can be also defined by the context in which deception takes place.

Deception is a very comprehensive and high level category but, in many fields of study, the link between this general concept and its occurrences tends to be ignored. Consequently, analysis is focused on a few sub categories of specific interest: semblance is particularly studied in semiotics and literature, cheating in law and so on. In our opinion, a strong and general notion of deception is necessary. A general notion allows different concepts to be organised within single theory. A strong notion highlights the boundary between deception and apparently similar activities. (For example, irony is not truly deceptive: the ironic message, though literally false, is usually uttered in a way that lets the recipient know that the speaker does not believe what they are saying.)

4.1.1. Cheating and fraud

We claim that a particular deceptive activity (cheating) necessarily occurs in exchange relationships. However, cheating is not the only sub category of deception performed in exchange relationships. In practice, it is rather close to fraud even if there are some important differences between these two concepts.

- Fraud is a deceptive activity which aims to induce someone into a relationship of exchange or co-operation which they would not enter into if they knew the true state of affairs. By contrast, cheating occurs after the relationship has begun, when an agent breaks a commitment. Obviously, the information denied in fraud has to be necessary to the agent's decision to participate in the relationship.

Fraud is particularly common in exchange (because this social relationship encourages deception). However, it may occur in any situation in which an agreement must be reached, whereas cheating is specific to the context of a particular prior agreement.

- Furthermore, cheating is never benevolent (the cheat never has good intentions). A cheat is violating an agreement that has already been drawn up and is deceiving any partners involved. There could be other ways to change the situation without committing fraud. For example, even if an agent is confident it is giving something *better* to its partner, it still has a duty to inform the partner *in* order to avoid cheating. The only one way to break the commitment without being guilty is to reach a new agreement.

On the contrary, fraud can be benevolent (usually when friendship is involved). No fraud that also implies cheating can be considered benevolent.

- The agent who is being cheated obtains something worse, less, different (perhaps better) than what the counterpart promised. Fraud does not necessarily implies a loss in economic terms for *D* and a profit for *L* although this is frequently the case. Firstly, fraud can induce *D* to an (unintentionally) favourable agreement. Secondly, *L*'s goal (to obtain a benefit at the expense of *D*) might not be satisfied. (This could occur as a result of a mistake by *L*.)
- Finally, fraud can take place without cheating and vice versa. However, neither can occur without deception. An apparently non-deceptive fraud is actually either a bribe or an exercise of monopoly power. (When *L* deliberately creates a shortage of a good, *D* has no choice but is aware of *L*'s aggression.)

A non-deceptive cheat is a simple agreement violation and it is usually arises from the contract. Cheating occurs when aggression on two different levels occurs at the same time. Firstly, *L* violates an obligation. Secondly, *L* deceives the other agent (or whoever is in control) by letting them believe that the obligation has been fulfilled. This dualism of cheating and fraud differentiates them from simple mistakes. When *L* gives something different to *D* (either in quality or quantity), the obligation has not been objectively fulfilled, but the mental elements of deception are missing. There is no intention to make the counterpart believe something different from what *L* believes.

4.2. THE ROLE OF DECEPTION IN PRO-SOCIAL ACTION

Deception is aggressive because it aims to deny another agent information it would like to have.

- *D* could decide to make a different decision knowing the truth
- *D*'s general goal of not being "taken for a ride" is compromised
- There is a social law about sharing knowledge and not giving false information. If we ask for the way to the railway station in a foreign town, people typically show us the right way if they know it. Deception breaks this social rule.

An agent's motivations may not define its behaviour. The aggression involved in deception may result from the benevolent aim of fulfilling some other (more important) goal. Consider the mother role. It is accepted (under some circumstances) that parents may know the interests of a child better than it does. A mother can decide to deceive her child in order to help him reach a higher level goal. She can lie about what is on TV to induce the child to do his homework.

Therefore, fraudulent behaviour can be benevolent. It is interesting that a really pro-social action leaves very little room for deception because there is no compelling stimulus to deceive when people are trying to reach a shared goal. Fraudulent activity has no close relationship with co-operation because the structure of co-operation prevents deception. By contrast, exchange fulfils private needs and permits many forms of deception for this reason.

4.2.1. The role of deception in co-operation

Co-operation is a relatively *safe* relationship. There are structural reasons why it tends to restrict the possibilities and, consequently, the agents' will to deceive each other. On the contrary, the exchange relationship necessarily leads to fraudulent behaviour. In the following paragraphs, we will analyse the differences between exchange and co-operation and show how these differences are linked to deception.

Co-operation is defined by a shared goal among interdependent agents. In other words, if one party's goal is reached, the other's is also achieved. Both agents must perform a plan for the goal to be fulfilled. Each agent is interested in complete and adequate performance of the actions in the plan leading to the shared goal. (This interest, and mutual knowledge of it, is often a necessary condition of a successful shared plan.)

Obviously, a cunning agent could always free ride, but this is far from easy in practice. Firstly, because a range of competencies are necessary. In this case, nobody can substitute for the fraudulent agent. Secondly, the means of reaching the shared goal are often close to each other and, in consequence, it is easier to control the counterpart's actions. For the same reason, continuous monitoring of the partner's behaviour is also quite economical. Even when the competencies and actions involved in the plan are identical, they are still necessary to accomplishment of the goal. Suppose your car suddenly dies half way up a hill. While pushing it forward, I can simply *pretend* to apply my strength, but instead let you do most of the pushing. In this case, the agents' goals appear to be shared but they are not. Indeed *L* pretends to share a goal with *D* for the purposes of exploitation. *L* thus deceives *D* about the relationship between their goals.

In this situation, unfulfilled obligation is more likely to be discovered immediately, the deceiver punished and the collaboration dissolved before a thorough loss occurs. This is a very important point. In exchange, *D* (the buyer/recipient) cannot always succeed in controlling goal fulfilment even at the end of the plan because of a lack of competence to check *L*'s actions (Brockway 1993, Holley 1998).

In sum, co-operation can be considered as a relatively deceit-free relationship for two main reasons. Firstly, its structure prevents attempts at free riding. Secondly, it allows partners to exercise mutual control at reduced cost.

4.2.2. The role of deception in exchange

The exchange relationship has a particular tendency towards deceptive activities. Unlike co-operation, exchange has a loose structure: agents have only one goal in common, the wish to exchange in order to reach their individual goals.

- Each agent has a private goal and is interested only in its fulfilment. The goal achievement of the other agent is only a means to an individual end. Furthermore, there is a gap between satisfying objective and subjective commitments. To have a goal adopted, *L* only needs to make *D believe* that its goal has been achieved.
- To make matters worse, there is low probability of controlling the other agent's actions during the process. Such control is both difficult and expensive.

  The exchange process involves six steps (see below). One of them, bargaining, is explicitly characterised by deception. In fact, lying is a fundamental element of exchange. Without deception - if people immediately announce the price at which they actually want to buy or sell - this step of the exchange process collapses. No bargaining occurs and agreement is suddenly reached.
- Another property of exchange that incites agents to fraudulent behaviour, especially in commercial transactions, is the natural informational

asymmetry between the buyer and the seller. This is particularly strong as regards the quality of goods. One agent is more informed than the other and has more opportunities for deception. By contrast, barter is symmetrical because each agent has more information about its own good.

- In the game theory literature (Muthoo 1999, Vulkan and Jennings 1999), non simultaneous transactions are considered a major source of cheating and fraud. If non simultaneous transactions favour deception, simultaneous ones do not eliminate it (This is obviously true of counterfeit money. Moreover, the buyer often has no means to control every feature of the good purchased.)
- Of course, cultural and moral issues are relevant in this context. It has been experimentally demonstrated (Lewiki 1983) that people believe fraud is more likely to occur in exchange than in other interactions. In order for exchange to be possible, a strong morality is crucial. Trust in the partner, in the authority and in the rules of society are both necessary to and (at the same time) potentially eroded by this social relationship.

4.2.3. Deception in the exchange process

As we have seen, exchange is closely tied to deception. Moreover, in this social relationship, deception is never benevolent. It is carried out for one's own interest rather than for any social benefit.

We can observe different kinds of deception at each step in exchange. The exchange process can be divided into seven steps. Five of these lead up to agreement: *identification of need, search for a partner, checking the necessary conditions, choice of partner, bargaining* and *agreement*. Two occur immediately after agreement: *transaction* and *post transaction*. Before agreement, there can be fraud. After the transaction, there is room for cheating.

At each step, agents can use several fraudulent strategies. Some are more appropriate than others in specific conditions. In order to provide examples of fraud and cheating, we have applied the taxonomy of deception strategies introduced by Castelfranchi and Poggi (Castelfranchi and Poggi 1998). This includes

- Omission of important information (closely linked to the problem of information disclosure in sales)
- Concealment (hiding important information)
- Falsification (making *D* believe something different from *L*)
- Confirmation of a false belief
- Negation of a true belief
- Falsification intending to conceal something

There are several objects about which *L* can practice deception during exchange:

- *Needs*: *L* can lie about its own needs or about those of *D*
- *Good:* *L* can simulate the possession of the *merchandise* and/or its characteristics
- *Real intentions*: *L* can practice deception about what it intends.

Let us consider each step in some detail.

Identification of needs. The first step of the exchange process is the identification of a need. An agent realises that it cannot achieve one or more of its goals. A need implies that an agent believes a given entity (action or resource) is a means to achieve some goal. As a consequence, the agent forms a sub goal to obtain that entity. As well known research has shown (Ekman 1985, Elster 1989, Sauerman 1978) the agent can practice deception about the means to satisfy a particular goal and about the goal itself. In order to compensate the partner as little as possible, an agent can deny or minimise its need. Alternatively, the partner can be deceived about its need. It may be induced to believe that *L* controls the best (or only) means to reach a particular goal. Failing this, *L* may be persuaded to believe that it is necessary to perform an action (such as a legal obligation) involving a given means (*L* can really experience that deceptive need in consequence of a "fictitious" goal provided it has been internalised.) Obviously, in these cases, communicative strategies are more effective than strategies of omission.

Search for a partner. Here, fraud relies on the possibility of finding alternative partners (alternative transactions) and on the possession of goods. *L* can simulate the possession of a given resource to induce *D* to exchange. This kind of fraud inevitably implies cheating because L will not fulfil the commitment.

Checking the necessary conditions. *L* can deceive by simulating a trustworthy attitude and dissimulating a conflict of interest with the partner. *L* objectively has the resource needed to make the transaction possible but, for some reason, the goals held by *L* and *D* are incompatible. In this situation, *L* can promise to reciprocate, achieve the goal and break the promise soon after.

Choice of Partner. At the fourth step, there is no new opportunity for fraud. However, many lies planned in the previous stages are now realised. *D* now believes in *L*'s trustworthy attitude, ability and will to reciprocate and the suitability of *L*'s resource. This is a brief stage but it represents a first important reduction in the set of alternatives. In principle, the process is still reversible at this point but to go back can be very expensive. The chosen partner is therefore highly likely to become the counterpart in exchange.

Bargaining. Contrary to the usual emphasis, bargaining is just one part of the process albeit a very important one. At this stage, agents lie about prices for buying and selling and about the quality and quantity of their goods (Akerlof 1970, Binmore and Dasgupta 1987).

Bargaining is characterised by lying. Without some form of deception it is not possible to haggle. When agents declare their prices, they decide whether a favourable transaction can be performed (Castelfranchi 1999).

Transaction. After agreement is reached, transaction takes place. Now agents can cheat by delivering something different from what they promised. Sometimes, cheating is associated with an intention not to be discovered. Often, cheating is inevitably discovered during the seventh stage (post transaction). In spite of this, the cheat can still avoid punishment especially when they have negotiation power (cf. Conte et al. 1998).[6]

As we can see, deception comes into its own in a relationship like exchange. Open aggression towards the counterpart usually does not favour a transaction (depending upon the aggressor's social, physical and economic power). But unconditional adoption of another's goal is unnatural too. The trade-off between self-interest and the need for help from others leads to a sophisticated form of communication in exchange processes. Lies, half-truths and concealment all derive from the complexity and internal dilemmas of this social relationship.

5. The Utility of the Present Analysis

Why do we need to predict and account for the occurrence of deception and fraud?

5.1. SCIENTIFIC REASONS

Since its early days, social theory has investigated whether (and to what extent) agents can achieve global benefits spontaneously. The social scientific community is notoriously split in two major parties: the bottom up ("invisible hand") party and the top down ("deliberate design") party. The gap between these parties arises from this question. Members of the former party argue that agents can spontaneously achieve global interests while members of the latter party argue that they cannot.

The gap between these parties is actually smaller than appears to be. The

[6] This implies the power to choose which partners to make transactions with.

latter party believes in the necessity of designed solutions, policies, norms, institutions and so on. The former party also believes in the evolution of spontaneous institutions (norms and conventions) to reduce defection: agents defeat one another unless and until social norms evolve. Therefore, both parties are fundamentally Hobbesian. Both assume a natural wildness in social agents, and believe norms and morals are necessary to avoid social ruin. However plausible this may appear, it is a pre-established and unilateral view of sociality. A more careful investigation highlights the variety of social action and attenuates cynicism.

5.2. APPLICATION-ORIENTED REASONS

There are also more practical reasons for interest in this issue. In particular, a theory about the role of deception in social action may help predict (and even possibly manage) social difficulties and disturbances in different domains:

- Solidarity networks: Which type of effects will follow from which type of interaction? Clearly defection is a disturbing factor which endangers the stability of social relationships, favours migration, hinders integration and so on. In multicultural societies, for example, the role of such a theory is crucial. It might provide predictions, proposals for regulation and heuristic hints to various kinds of policy making.
- Organisational structures: What kind of interaction should be promoted to increase functionality, productivity and stability in organisational structures? Which types of social relationship allow agents to engage in effective interaction? How can we reduce the frequency of defection?
- Innovation: Which kind of structural relationships are favoured in given environments and by given conditions? What type of social action would we expect to prevail in, say, agent mediated interaction? Which type of social action can we expect to be favoured by electronic interaction? This question points to the problems that information societies might pose and, consequently, helps us to envisage and design possible solutions.

Any of these questions can fruitfully be addressed within a social simulation framework. Indeed, this framework lends itself not only to the provision of useful techniques but also to models, operational hypotheses and notions about crucial social scientific issues.

5.2.1. Solidarity networks

There is a great deal of simulation modelling about issues of solidarity. (For both a review and an original research contribution see Hegselmann and Flache 1998.) What is still needed is a study of the respective effects of co-operation and exchange on the maintenance of solidarity. Two main questions should be addressed:

- What are the structural conditions for the emergence of different types of social action? When would we expect to observe co-operation and when exchange?
- What are the respective effects of different social actions on the nature of future relationships, on group stability and so on?

Conditions. Many simulation studies have been conducted on the emergence of dependence networks among heterogeneous agents. (See Conte and Sichman 1995, Conte and Castelfranchi 1996, Conte *et al.* 1998). However, these simulations were not meant to assess (by "reverse engineering") which conditions facilitate co-operation rather than exchange or vice versa. Instead, the model presented in this paper could be form the basis for a simulation to explore the conditions which favour the occurrence of cooperation and exchange.

It can be hypothesised that, other things being equal, the probability that two or more agents will co-operate is lower than the probability that they will exchange. In fact, co-operation requires:

- complementary actions
- goal sharing
- belief sharing about
  - plans: beliefs about the shared goal and how to achieve it
  - the complementary nature of agents

By contrast exchange only requires:

- the existence of plans allowing two or more actions to be performed for a single goal. Thus exchange is possible in much simpler contexts.

In exchange, agents need to be interdependent and share social beliefs (about their reciprocity).

Are there conditions that compensate for this inherent asymmetry? In part, these conditions have been analysed by game theory. For example, the difficulty of cheating which arises from the frequency of repeated encounters and so on. This condition serves as an obstacle to exchange more than to co-operation. Others conditions (like emergencies) have not been addressed. Sudden environmental changes may call for extraordinary co-operation because they increase the number and importance of shared goals among interdependent agents. (Events like predators and famine move us away from mere "preference" and towards survival goals shared by a wider humanity.)

Effects. What are the effects of co-operation and exchange on future relationships? This is an interesting and complex question that must be analysed from different perspectives. There are two variables along which effects of social action are usually measured and compared at the population or group level. These can be easily investigated through simulation:

- Efficiency: the global benefit achieved in each condition. In game theory, it is assumed by definition that the co-operative moves will produce the highest global benefit. This is also confirmed by our subsequent analysis.
- Equity: the deviation between individual outcomes and average (variance). Again, as game theorists argue, co-operation is more equitable than exchange. (See also Conte and Castelfranchi 1995). But, at least according to the "invisible hand" mentality, we pay dearly for equity in terms of efficiency since an equal distribution of resources may depress (economic) growth. However, simulation results show (Epstein and Axtell 1996) that a high level of inequality caused by trade leads to an impoverishment of the global population. (See also Paolucci et al. 2000.)

	Efficiency	**Equity**
Co-operation	+	+
Exchange	+	-

There are other measures by which the two phenomena could be compared:

- Compatibility with intrinsic motivations: The goal of exchange is extrinsic or instrumental to a further benefit. This is not the natural but the social (or artificial) consequence of goal adoption. Reciprocation is a compensation for goal adoption. By contrast, in co-operation, the achievement of a shared goal is a natural consequence of the participants' activity. The common plan is a natural means to achieve the goal in question. Hence, the goal of co-operation is intrinsic. For example, positive incentives have been found (Lepper forthcoming) to reduce intrinsic motivations to action (This mechanism is known as over-justification.) Therefore, exchange is not recommended when agents are expected to do something spontaneously. (For example, in educational contexts, contracts and transactions should be applied in moderation.)
- Satisfactory Outcomes: The consequences of exchange are not necessarily as satisfactory at the individual level as the consequences of co-operation. Agents may realise they have been cheated by their partners. In consequence, their goals will be achieved only in part or at a higher cost than expected. Beside the objective damage, agents may bear the psychological costs of an unfair deal: loss of self-esteem or reputation, insecurity, helplessness, social withdrawal, mistrust and so on.

- Stability in relationships: Unfair deals may endanger social relationships. As soon as agents realise to have been cheated or deceived, they look for other partners or may even give up their goals. On the other hand, cheats may be compelled to find fresh victims to exploit. This may represent a serious obstacle to social efficiency and may lead to impoverishment and stagnation.[7]
- Group cohesiveness: What are the effects of exchange interactions among group members on the future cohesion of the group? Popular wisdom points to this phenomenon in recommending that one should never do business with friends. Notably, group cohesion correlates positively with important social phenomena, for example the reaction of bystanders to emergencies (cf. Latané 1981). The more cohesive the group of bystanders, the higher the probability, timeliness and efficacy of intervention in emergencies.
- Migration and segregation: A further dynamic consequence of cheating in exchange is migration. Agents may withdraw from a potential "market" (aggregate of agents seeking partners for exchange) or try to move to a new market (migrate). Finally, they may stick to a sub group and exchange only within the boundaries of that group. What are the specific effects of these options? How does co-operation in its turn affect social and spatial dynamics, the processes of migration and segregation?

5.2.2. Organisational structure

An interesting domain of application for the present analysis is the organisational context. Which conditions increase functionality, productivity and stability in organisational structures? Which social relationships allow agents to engage in effective interaction? How can we reduce the frequency of defection?

This question is relevant for horizontal organisations (Business Week 1993). Essentially, horizontal organisations are flexible social entities in which all the sub parts involved hold direct or indirect exchange relationships with external entities (other firms or companies) and respond directly to them. Horizontal organisations have greater transparency and more distributed responsibility than traditional pyramidal organisations. At the same time, the occurrence of co-operation is reduced to the benefit of exchange. To what extent does a horizontal (as opposed to a pyramidal) organisation allow for deception and fraud both within and without the organisation? And what are the costs of reduced co-operation? How stable and cohesive are horizontal organisations

[7] This effect is quite simple to check using simulation. Agents can be made to keep track of previous deals and refuse interactions with cheats (unless the importance of the goals for which they seek help is greater than the predictable costs of being cheated). The effects of exchange and co-operation can then be compared in different populations and within the same population.

over time? To answer these questions, it is necessary to investigate the cognitive ingredients and effects of the phenomena under investigation.

5.2.3. Innovation

Types of social action and the differential incidence of fraud may also affect innovative technological applications. For example, which type of social action can we expect to prevail in electronic interactions? Will the massive use of the Internet promote co-operation or exchange? Will the information society be a co-operative and pro-social society or a bargaining and deceptive one? The Internet (*Communications of the ACM* 1994) was expected to increase the extent of participatory and communitarian processes in current societies. In fact, it is mainly used for market activity (e-commerce). This could have been predicted in terms of the present analysis. Co-operation is less likely in larger networks (where conditions for co-operation are less likely to occur) and in indirect relationships (where deception is easier and reciprocal control among partners is more difficult).

An interesting question for simulation modelling in this context is what effects fraud and deception in exchange will have for the future developments of the electronic market. For example, what are the long-term effects of the spread of (bad) reputation? That reputation may interfere with the search for a partner has already been shown (Crabtree 1998). An artificial system with a bad reputation (such as Bargain Finder) may be mistrusted because of its negative social effects. The question is which kinds of deception and fraud can be tolerated in exchange and which cannot? Which specific kind of deception and cheating gives rise to a bad reputation? Which types of cheating are likely to cause an essential impairment of the exchange activity itself by neutralising or reducing (economic) efficiency? All these questions can be addressed by simulation models provided that the cognitive ingredients of social action are modelled as we have endeavoured to do throughout this paper.

References

Akerlof, A. (1970). The market for lemons: Quality uncertainty and the market mechanism. In: *Quarterly Journal of Economics*, 84(3), 488-500.

Axelrod, R. (1984). *The Evolution of Co-operation*. Penguin Books, Harmonsdsworth.

Axelrod, R. (1986). An evolutionary approach to norms. In: *American Political Science Review*, 4, 1093-1111.

Axelrod, R. (1997). *The complexity of cooperation*. Princeton University Press, Princeton, NJ.

Axelrod, R. and Dion, D. (1988), The further evolution of cooperation. In: *Nature*, 242, 1395-1390.

Bicchieri, C. (1990). Norms of cooperation. In: *Ethics*, 100, 838-861.

Binmore, K. (1987). Modelling rational players I. In: *Economics and Philosophy*, 3, 9-55.

Binmore, K. (1994). *Game Theory and the Social Contract: Playing Fair.* MIT Press, Cambridge, MA.

Binmore K. and Dasgupta, P. (eds.) (1987). *The Economics of Bargaining.* Basil Blackwell, Oxford.

Binmore, K. and Vulkan, N. (1997). Applying game theory to automated negotiation, <http://else.econ.ucl.ac.uk/papers.htm>.

Brockway, G. (1993). Limited paternalism and the salesperson: A reconsideration. In: *Journal of Business Ethics*, 12, 275-279.

Business Week (1993). *The Horizontal Organisation.* 20 December 1993.

Castelfranchi, C. (1999). Lies in bargaining. In: *Ethics and Information Technology*, in press.

Castelfranchi, C. and Poggi, I. (1998). *Bugie Finzioni e Sotterfugi: Per una scienza dell'inganno.* Carocci Editore, Roma.

Conte, R. (1999). Social intelligence among autonomous agents. In: *Computational and Mathematical Organisation Theory,* 5(3), 203-228.

Conte, R., and Castelfranchi, C. (1995). *Cognitive and Social Action.* UCL Press, London.

Conte, R., and Castelfranchi, C. (1996). Simulating interdependence networks: A two-way approach to the micro-macro link. In: Troitzsch, K.G., Mueller, U., Gilbert, N. and Doran, J.E. (eds.) *Social Science Microsimulation.* Springer-Verlag, Berlin.

Conte, R. and Sichman, J.S. (1995). DEPNET: How to benefit from social dependence. *Journal of Mathematical Sociology*, 20(2-3), 161-177.

Conte, R., Castelfranchi, C. and Pedone, R. (1999). The impossibility of cooperation in PD-game. In: Meyer, J.-J. C. and Schobbens, P.-Y. (eds.), *Formal Models of Agents*, Springer-Verlag, Berlin.

Conte, R., Castelfranchi, C. and Veneziano, V. (1998). The computer simulation of partnership formation. *Computational and Mathematical Organization Theory*, 4(4), 293-315.

Crabtree, B. (1998). What chance software agents. In: *The Knowledge Engineering Review*, 13, 131-137.

Ekman, P. (1985). *Telling Lies: Clues to Deceit in the Marketplace, Politics and Marriage.* W. W. Norton and Company, New York.

Elster, J. (1989). *Nuts and Bolts for the Social Sciences.* Cambridge University Press, Cambridge.

Epstein, J.M. and Axtell, R. (1996) *Growing Artificial Societies: Social Science from the Bottom Up.* MIT Press, Cambridge, MA.

Gauthier, D. (1986). *Morals by Agreement.* Clarendon Press, Oxford.

Gauthier, D. (1993). Uniting separate persons. In: Gauthier, D. and Sugden, R. (eds.), *Rationality, Justice and the Social Contract.* Harvester Wheatsheaf, Hemel Hempstead.

Heckathorn, D. (1988). Collective sanctions and the creation of prisoner's dilemma norms. In: *American Journal of Sociology*, 94(3), 535-562.

Hegselmann, R. and Flache, A. (1998). Understanding Complex Social Dynamics: A Plea For Cellular Automata Based Modelling. In: *JASSS - Journal of Artificial Societies and Social Simulation*, 1(3), <http://www.soc.surrey.ac.uk/JASSS/1/3/1.html>.

Hirshleifer, J., and Coll, J.C.M. (1988). What strategies can support the evolutionary emergence of cooperation? In: *Journal of Conflict Resolution*, 2, 367-398.

Holley, D.M. (1998). Information disclosure in sales. In: *Journal of Business Ethics*, 17, 631-641.

Howard, N. (1971). *Paradoxes of Rationality: Theory of Metagames and Political Behavior.* MIT Press, Cambridge, MA.

Latané, B. (1981). The psychology of social impact. In: *American Psychologist*, 36, 343-356.

Lepper, M.R. (forthcoming). Theory by numbers? Some concerns about meta-analysis. In: *Applied Cognitive Psychology.*

Lewiki, R.J. (1983). Lying and deception: A behavioural model. In: Bazerman, M.H. and Lewiki, R.J. (eds.), *Negotiating in Organisations*, Sage Publications, Beverly Hills, CA.

Lomborg, B. (1994). Game theory vs multiple agents: The iterated prisoner's dilemma. In: Castelfranchi, C. and Werner, E. (eds.), *Decentralized A.I. 4.* Springer-Verlag, Berlin.

Luce, R.D. and Raiffa, H. (1989). *Games and Decisions.* Dover, New York.

Macy, M. (1998). Social order in artificial worlds. In: *JASSS - Journal of Artificial Societies and Social Simulation*, 1(1), <http://jasss.soc.surrey.ac.uk/JASSS/1/1/4.html>.

Maynard Smith, J. (1982). *Evolution and the Theory of Games.* Cambridge University Press, Cambridge.

Miller, G., Galanter, E. and Pribram, K.H (1960). *Plans and the Structure of Behavior.* Holt, Rinehart and Winston, New York.

Molander, P. (1985). The optimal level of generosity in a selfish, uncertain environment. In: *Journal of Conflict Resolution*, 4, 11-18.

Mueller, U. (1987). Optimal retaliation for optimal cooperation. In: *Journal of Conflict Resolution*, 4, 692-724.

Muthoo, A. (1999). *Bargaining Theory with Applications.* Cambridge University Press, Cambridge.

Paolucci, M., Marsero, M. and Conte, R. (2000) What's the use of gossip? A sensitivity analysis of the spread of respectful reputation. In: Suleiman, R., Troitzsch, K.G. and Gilbert, G.N. (eds.) *Tools and Techniques for Social Science Simulation.* Physica-Verlag, Heidelberg.

Rasmusen, E. (1990). *Games and Information: An Introduction to Game Theory.* Basil Blackwell, Oxford.

Rubinstein, A. (1986). Finite automata play the repeated prisoner's dilemma. In: *Econometrica*, 39, 83-96.

Sauerman, H. (1978). *Bargaining Behaviour.* J.C.B. Mohr (Paul Siebeck), Tübingen.

Ullman-Margalit, E. (1977). *The Emergence of Norms.* Oxford University Press, New York.

Vulkan, N. and Jennings, N.R. (1999). Efficient mechanisms for the supply of services in multi-agent environments, <http://else.econ.ucl.ac.uk/papers.htm>.

HUNTING THE UNICORN

An exploration of the simulation of small group leadership[a]

NIGEL GILBERT
Centre for Research on Simulation in the Social Sciences
Department of Sociology
University of Surrey
Guildford, United Kingdom

and

EDMUND CHATTOE
Department of Sociology
University of Oxford
Oxford, United Kingdom

Abstract

The chapter explores whether social simulation research has to adopt a realist epistemology, or whether it can operate within a social constructionist framework. Research on leadership in small groups is used to illustrate the argument. The main question to be explored is how one can understand the emergence of identifiable leaders from interaction in small groups. The focus is on the emergence of 'macro' phenomena (in this case, leadership and follower roles from an undifferentiated collection of agents) as a result of interactions between agents at the 'micro' level.

The chapter outlines the history of leadership studies and argues that the changing approaches to the topic can be associated with changing epistemological orientations. It is shown how each of the approaches to leadership could have been supported by simulations, and thus how the different approaches to social theory generate different research questions and different simulations.

> A man breaking his journey between one place and another at a third place of no name, character, population or significance, sees a unicorn cross his path and disappear. That in itself is startling, but there are precedents for mystical encounters of various kinds, or to be less extreme, a choice of persuasions to put it down to fancy; until – "My God," says a second man, "I must be dreaming, I

[a] Acknowledgement: This chapter is based on a conference presentation at the SimSoc workshop at Cortona, Italy, 20-21 October 1997. We gratefully acknowledge the comments of participants at the Workshop.

N.J. Saam and B. Schmidt (eds.)
Cooperative Agents Applications in the Social Sciences, 109-124

> thought I saw a unicorn." At which point, a dimension is added that makes the experience as alarming as it will ever be. A third witness, you understand, adds no further dimension but only spreads it thinner, and a fourth thinner still, and the more witnesses there are the thinner it gets and the more reasonable it becomes until it is as thin as reality, the name we give to the common experience... "Look, look" recites the crowd. "A horse with an arrow in its forehead! It must have been mistaken for a deer." (Stoppard 1968: 14-15)

Those developing simulations of social phenomena have largely confined themselves to representing a quite narrow range of sociological theories, ones that are broadly realist. This chapter is an attempt to identify the issues that would need to be addressed in order to widen the scope of simulations to encompass other approaches to social theorising. It focuses particularly on the requirements of social constructionist theories which are probably more common in current sociology than realist ones are. Much the same is increasingly the case in social psychology.

The chapter explores whether it is unavoidable that social simulation research should adopt a realist epistemology.[1] In order to illustrate the argument, we shall use a running example: research on leadership in small groups. This topic interests both social psychologists and organisational theorists. It has had a very long history, going back to the early 1940s, although approaches have changed radically since then. The main question that we shall explore is how one can understand the emergence of identifiable leaders from interaction in small groups. The focus will therefore be on one of the main themes of recent simulation research: the emergence of 'macro' phenomena (in this case, leadership and follower roles from an undifferentiated collection of agents) as a result of interactions between agents at the 'micro' level. Other work has also been interested in these questions, for example, the EOS series of experiments which explored mechanisms for the emergence of hierarchies using theories developed to explain the emergence of social complexity in Palaeolithic south-west France (Doran 2000).

We shall outline, very briefly, the history of leadership studies and argue that the changing approaches to the topic can be associated with changing epistemological orientations. We shall also indicate how simulation could have supported each of these approaches to leadership, although as far as we know no simulation work has ever done in this field. This will allow us to show how the different approaches to social theory generate different research questions and thus different simulations. In turn, these simulations raise important points about leadership. The chapter thus interweaves three strands: the philosophy of social science, developments in leadership studies and simulation as a method.

[1] We thank Petra Ahrweiler for raising this issue originally at the Schloß Dagstuhl Workshop on Social Science Microsimulation: Tools for Modelling, Parameter Optimization and Sensitivity Analysis, 5-9 May 1997.

1. Theories of leadership

1.1. THE TRAIT APPROACH

The development of leadership studies has been very well summarised by Alan Bryman (1996, 276-291). He identifies four stages of leadership research, very roughly aligned with the decades between 1940 and 1980. At the end of each stage, the emphasis of research changed. While work in the previous traditions continued, it made progressively less impression. The first stage, flourishing until the late 1940s, he calls the trait approach. This sought to identify the particular qualities and characteristics of leaders, on the assumption that such qualities were unevenly distributed and that those who possessed them became the leaders of small groups. In a nutshell, this approach assumed that leaders were born not made. The associated research consisted of applying a range of personality inventories to identify those traits that were most associated with effective leadership.

Bryman notes that a host of different traits were examined, including physical qualities (such as height and appearance), abilities (such as intelligence), and personality characteristics (such as conservatism, self-confidence and extroversion). Unfortunately, the results of these studies tended to show rather weak associations and different studies contradicted one another in identifying particular traits as important. For this reason, the trait approach fell into disfavour.

The lack of clear results is only one problem with this research. Another is the unsatisfactory nature of its methodology. It seeks to identify an empirical regularity: that some particular trait is associated with 'leadership', but without bothering much about why the trait and leadership are linked. We can see such studies as characteristic of a standard (and rather unsophisticated) variety of positivism, where the object is to generate an explanation by combining a law and a set of initial conditions. In this case, the approach sought to induce a 'law of leadership', for example, that individuals well endowed with certain attributes make good leaders. As Halfpenny (1997) notes, positivism can be divided into conventionalism and empiricism. This distinction is characterised neatly in a quote from Crutchfield (1992) suggesting that the distinction between science and engineering (or between conventionalism and empiricism in this context) is that

> the scientist ... presumes to be focused on what the model means vis-à-vis natural laws. The engineering view of science is that it is merely data compression.

What would a simulation intended to illustrate and develop a positivist theory of leadership, such as the trait approach, look like? It is not hard to imagine it. We would need a collection of agents each possessing attributes representing the traits supposedly associated with leadership. These traits could be 'built in' to the design of the agents. The simulation would require the agents to come to a consensus. We would then look to see whether the agents which were central to the interaction, or which influenced the others, that is the leaders, were also those who had the expected traits.

Consider agents who have either positive or negative opinions about some issue. These agents are placed in a setting where they all discuss the issue with a view to influencing each other. The stopping condition is that they are all of the same opinion, either all positive or all negative. A very simple model of this setting would have agents 'discussing' their opinions with every other agent. Let us assume as the 'null' case that all agents are identical in every respect and that the result of every discussion between two agents is that both come to agree. However, whether the agreement follows the opinion of one or the other is a matter of chance. The consequence is that, after many rounds, the opinion of the whole group will reach consensus. However there is an equal likelihood of obtaining a positive or negative opinion consensus. Now let us introduce a 'trait' that is differentially distributed through the group. This might represent persuasiveness, authority or just strength of opinion. In this new case, if two agents with different opinions discuss and reach an agreement, the chance of one influencing the other is determined by the relative magnitudes of the agents' traits. The result is that positive and negative opinion consensus are no longer equally likely, Instead, the type of consensus is completely determined by the opinion of the agent that has the trait of the largest magnitude, provided that sufficient rounds of discussion are allowed to elapse.

One value of this simulation is that it immediately raises the question of what we mean by 'leadership'. This question troubled the early researchers as well. One of the enduring difficulties in the field has been a lack of consensus about how leadership can be recognised. Bryman asserts that a definition of leadership (originally formulated by Stoghill in 1950) is one that most researchers would accept:

> Leadership may be considered as a process (act) of influencing the activities of an organised group in its efforts towards goal setting and goal achievement ... (Stoghill 1950)

From the perspective of simulation, however, this definition is woefully inadequate. A 'process' (a causally connected sequence of events) is not an 'act' (an intentional piece of behaviour). What precisely does 'influencing' mean? Is it not tautologous to define leadership in terms of an organised group, when the leader is presumably the source of the organisation? Whose 'efforts' are being referred to? Is it possible to recognise the 'efforts' of a group? And to whose goals are the efforts directed? The definition seems to assume common goals, but how does goal consensus arise? These are not pedantic points: they all need to be clarified before one could build a computational model in which this sense of leadership could be identified.

The primary value of simulation in this context is precisely to pose these kinds of definitional questions and to force clear answers. Once these issues have been settled, we can then explore the circumstances in which leadership emerges, given a distribution of traits among the agents. The aim could be to discover 'natural laws' about leadership that could be applied to all societies, artificial or human (see, for example, Doran 1997). Alternatively, the simulation could be used to make predictions about the emergence of leadership (in particular contexts such as a jury or an aircraft cockpit) given some set of initial conditions.

The former approach corresponds to Halfpenny's conventionalist version of positivism and the latter to his empiricist version. Conventionalist approaches to social simulation are not common, but empiricist simulations are. The most widely known examples are those micro-simulations used to make predictions about future household characteristics from data about a sample of current households. In this case, the concern is only with the empirical outcomes of the simulation and not with the status of the generalisations or laws which might be embodied in the code.

While one is concerned only with the 'engineering' approach, that is, with the predictions about the future which empiricist simulations aim to make available, the prime concern will be with predictive accuracy. However, as soon as one becomes interested in such issues as generalisation or induction, the way in which theory is translated into a model (and run as a simulation) becomes important. For example, one might speculate about the sorts of mechanism that could be built into the model to promote the emergence of leadership. However, as we move along this path, we seem to shift from a purely positivist approach to one which comes closer to realism. This can be illustrated by returning to the history of leadership studies.

1.2. THE STYLE AND CONTINGENCY APPROACHES

Bryman suggests that the second approach to leadership studies, following disillusionment with identifying leadership traits, was to focus on leadership behaviour, or 'style'. Instead of trying to identify the attributes of leadership, attention shifted towards distinguishing the personal styles of effective leaders. The implication was that, once identified, these styles could then be taught. In this view, leaders were made rather than born. Early work in the 1960s employed questionnaires (administered to subordinates) which consisted of batteries of items about their leaders. The responses were aggregated to form scores for a number of aspects of the leaders' behaviour. Two styles or behaviours were most investigated. Firstly, the so-called "consideration" component (behaviour reflecting the leader's concern about their subordinates as people, and the extent to which they respond to subordinates and promote camaraderie). Secondly, the "initiating structure" component (behaviour which involves defining closely and clearly what subordinates are supposed to do and how). Scores on these styles were then correlated with various measures of the group outcome such as performance and subordinates' job satisfaction.

Unfortunately, just as with the trait approach, these studies were, in Bryman's words, "... plagued with inconsistent results..." (p. 278). There were problems about imputing causality. For example, does behaving as a considerate leader cause an increase in subordinates' job satisfaction, or do satisfied subordinates enable leaders to be more considerate? In addition, the research procedure aggregated followers' view of leaders' behaviour, although different followers may have had different opinions. There were also difficulties with the scales used to measure behaviour. Furthermore, it became clear that the picture was rather more complex than a simple correlation between style and outcome might suggest. The link between style and performance was mediated by the context in which the leadership was exercised. For example, Fiedler (1967) proposed that task oriented leaders are most effective in high and low control situations, while relationship oriented leaders are best in moderate control situations, where 'control' measures how favourable the situation is to the leader.

One response to these difficulties was to emphasise the context specific nature of leadership, leading to a body of work classified by Bryman as the contingency approach. Both the style and contingency approaches inevitably led to speculation about the mechanisms that linked the adoption of certain leadership styles with particular outcomes. While certainly not departing far from the positivist approach described previously, some of the researchers began to think of their work more in realist terms. According to the realist position, social scientists should be concerned with uncovering the properties of the real causal mechanisms that operate in the world. (Scott 1995,173) Thus, the primary focus of realist science is on the mechanisms that are thought to underlie and generate the data that are observed. In terms of leadership studies, this obviously leads to an interest, not just in the supposed traits or styles of leadership, but to theories about how it is that particular behaviours, in particular contexts, give rise to effective leadership.

The realist epistemology is also a natural progenitor of computational simulations. It is a short step from developing *theoretical* models of mechanisms to developing *computational* models of mechanisms. It is also the case that simulations provide an appealing way of both formalising and exploring realist theoretical models. Nevertheless, we are not aware of any computer simulations which focus precisely on the issue of leadership, although the EOS family of models previously mentioned dealt with similar issues, and work by Opp *et al.* (1996) on revolutions has considered a related area.

Opp's model was constructed on the basis of rational choice theory. As an illustration of the type of realist model that it might be useful to explore in the context of leadership studies, let us therefore use rational choice to develop a simple leadership simulation. We might begin by asking why leaders occur at all. Note that this question is not one which naturally springs to mind when one starts from a positivist view of the attributes of leaders. Once one is concerned with models of mechanisms, however, it is almost the first question to consider.

Leaders (and followers), we might suppose, are of value when there are both advantages to engaging in collective action and where that collective action requires some organisation, for example, the division of labour and co-ordination of effort. Since individual agents aim to maximise their utility, when collective action could add to their utility, it is rational for them to band together into an organised group. However, leadership incurs costs for the leader (the time and effort required for the organisation, worry about whether the collective action is going to be effective and so on), although it may also bring rewards, such as power to control others, gratitude and perhaps even monetary benefits. If the costs for all agents in adopting a leadership role are greater than the individual benefits they expect to receive, collective action will not emerge. Every potential leader will wait for someone else to bear the costs, a version of the classic free rider problem. (Olson 1965)

On the other hand, if the benefits of leadership are perceived to be greater than the costs, everyone will try to lead, and all that will emerge is a battle over the leadership role, rather than a group with one leader and many followers. The way out of this paradox is to relax the implicit assumption that all agents are identical. If one agent perceives the benefits of leadership to be greater than the other agents do, or the costs to be less, or has some attributes (traits) which other agents consider advantageous in leadership, then the rational course of action is to fall in behind that agent.

In this way we can see that the kinds of traits and styles which were examined in previous leadership studies could have an important effect on the emergence of leadership, yet not necessarily be simply correlated with it. According to the mechanism we have sketched, any attribute of agents which is perceived to give an agent an advantage over others (including traits such as physique or assertiveness, and styles such as consideration and structure initiation) can trigger the emergence of that agent as the leader and the other agents as followers. Moreover, the mechanism also suggests why context, that is the situation in which the group is located, could be important to the emergence of leaders.

A computer simulation of agents forming groups with leaders could be constructed to explore this mechanism in more detail. First, one would need to specify some assumptions about the agents' motivation and possible actions. Let us consider one such design. The simulation would be based on the idea that agents are motivated to accumulate as much 'sugar' as possible and that sugar can be harvested most effectively by co-ordinated groups. (Epstein and Axtell 1996) The amount of sugar available to a co-ordinated group of n agents is β^n, where β is a constant. However, the group has to be managed by a leader and the cost of leadership is α. We assume that the sugar is distributed equally among the group except that that the leader is compensated for the effort of leading. Each agent has therefore to decide whether

(a) to go it alone, obtaining β^1 units of sugar;

(b) become a leader receiving $\dfrac{\beta^{n}-\alpha}{n}+\alpha$ units or

(c) become a follower receiving $\dfrac{\beta^{n}-\alpha}{n}$ units.

Each agent computes the respective potential gain from the choices available (join no group; join each potential group as a follower; create a group with a number of other agents as followers and self as leader) and selects the most favourable option (the one with the largest gain). However, since both the leader and follower choices require the co-operation of other agents (e.g. agent A can only lead agents B and C if B and C both conclude that joining this potential group as followers maximises their utility), there must be a sequence of negotiations[2] to establish the group, which continues until equilibrium is reached. Note, however, that a lot is assumed in the way the context is set up: for example, the formula for the sugar distribution assumes that there is, or could be, a leadership role and implicitly defines the leadership task as consisting of the co-ordination of group members and the division of the spoils among them.

1.3. THE NEW LEADERSHIP APPROACH

Once a leader has emerged, there must still be some mechanism for maintaining leadership over time. One way to ensure that followers continue to perceive an advantage in membership of the group is to give them rewards such as prestige or money for compliance with the leader's wishes. Another way is to use the leadership role to change the followers' goals to ensure that they continue to see the leader's goals as their own. These kinds of issues were addressed by research that Bryman classifies as the 'new leadership approach'. Various authors contrasted 'transactional leadership', an exchange between leader and follower in which the former offers rewards to maintain the follower's loyalty, with 'transforming leadership' in which the leader alters the aspirations of the followers so that the leader's and followers' goals are brought into harmony. Transformational leadership studies were popularised by management gurus such as Peters and Waterman (*In Search of Excellence*). This led to the current management emphasis on the value of articulating a leadership 'vision' and then persuading followers to share it through charisma.

These studies move away from the straightforward rational choice model towards a concern with the 'management of meaning'. Leaders engage with sense-making and interpretation for the group, aiming to create a consensus about what is real and what is important. It is still the case that the focus of this

---

[2] While these negotiations could be implemented using a formal negotiating procedure, such as contract net, there is no need to do so for the purposes of this simulation.

research is on the identification of real mechanisms through which leaders can achieve consensus over goals through processes of meaning management. However, instead of those mechanisms being theorised at the level of individual utility, they are considered in terms of symbols and values. While rational choice treats agents' utilities or preferences as exogenous to the theory, and considers only the benefits and costs evaluated in terms of those preferences, the new leadership approach suggests that leaders aim to manipulate agents' preferences. As such, the approach is not necessarily antithetical to rational choice, but supplementary. It may still be the case that followers weigh up the costs and benefits in a utilitarian fashion, but using evaluations that are influenced by the leader's management of meaning.

In contrast to the previous model, this approach would require us to explore ways of simulating the manipulation of some agent's goals by other agents. The previous model was set up with the simple assumption that all agents shared the same goal: to maximise the amount of sugar they received. The agents then mechanically computed the best group roles to achieve this, using a formula we provided. We now need to consider how the potential members of a group come to know what benefits accrue if they agree to join. This may involve processes of persuasion, argument and negotiation, at the symbolic or cognitive level, directed to convincing potential group members that the leader's goals should also be their goals. Once that had been achieved, the rational choice model may prove adequate for explaining why the members decide to follow the leader to obtain their common goals (rather than setting out alone, for example). If the environment changes (or is perceived to change), or members are persuaded that alternative goals are worth consideration, the leader will need to engage in persuasion once again. Thus from the point of view of simulation, goal-transformation is important because it brings in processes of negotiation and persuasion which are not a necessary part of the rational choice model.

In the same way that we did for the rational choice approach, we can sketch out a simulation in order to clarify the issues. For instance, while the previous model did not require the specification of any kind of internal agent architecture, other than the ability to calculate the potential payoff in 'sugar' for various strategies, a simulation of the goal transformation theory will certainly need agents with some (perhaps only primitive) cognitive capacity. One design would be for agents to have a single 'top-level' goal (to accumulate as much sugar as possible) and a planning mechanism which is able to decompose this top-level goal into tasks and sub-tasks. In the usual way of planners, each sub-task will have pre-conditions, post-conditions and actions associated with it (Fikes and Nilsson 1971). The agent will have a library of operators which provide templates showing it how to accomplish each task. An agent will then devise a suitable plan satisfying its top-level goal, using whichever operators it both has access to and are appropriate to the environment in which it finds itself. If there are two or more ways of achieving the same ends, it will choose the one which costs least in terms of resources. The agent's library of operators constitutes its knowledge of what it can do.

In this scenario, agent A can influence the action of agent B by providing the latter with additional knowledge, that is, with additional operators to add to its library. Provided that the new operator is useful, enabling agent B to satisfy its top-level goal more cheaply than plans which do not use the new operator, the agent will re-plan using that operator. If the new operator in fact involves, for example, following the lead of agent A, then this is what agent B will do. The communication of the operator from A to B has resulted in agent B being persuaded to follow agent A. Note that the transmission of the operator from A to B does not necessarily cause agent B to use that operator. It will only form part of a plan if agent B perceives it to produce a more efficient plan than alternative operators. Likewise, it is possible for agent B to receive operators from several other agents and to use just some of these operators, ignoring others. Potential inconsistencies between operators from different sources are not a problem since planning algorithms use only those operators which are expected to lead to the desired goal and do not even notice inconsistencies.

The rational choice simulation introduced above can be adapted to model this kind of approach. We would need to build a planner into the agents. Instead of giving all agents a simple formula for calculating the sugar gain if they joined groups of various sizes, they would instead possess operators which encapsulated the knowledge required for forming groups and calculating the expected outcome. For example, an agent would be primed with operators which had as a pre-condition the existence of a group of a certain size, with the agent itself as the leader, and as a post-condition the receipt of the appropriate amount of sugar.

This is only one rather simple design for a simulation that tries to model the idea of goal-transformation. We can easily imagine more complex variants. For example, the above discussion assumes that the operators communicated between agents are 'truthful': that is, an agent including an operator it received from another agent would actually obtain the amount of sugar indicated by the operator when the plan was executed. However, we can also build agents that lie, passing operators which overstate the sugar yield. Since the agents work in a situation where the truth of such forecasts cannot be determined before plan execution, they will not be able to recognise false operators for what they are. On the other hand, we could also allow agents to consider reputations (as measured by previous experiences of working together) in their choices about which operators to use.

The value of this simulation is that it begins to include communicative behaviour as a fundamental feature of the model. In contrast to the rational choice model, where co-operation between agents is 'accidental', in Conte and Castelfranchi's (1995, 144) classification, this model requires co-operation to be 'deliberate', that is, planned for in the 'minds' of the agents. We turn next to an approach for understanding leadership that focuses even more sharply on language and discourse.

1.4. ACCOUNTS OF LEADERSHIP

We now move beyond Bryman's four approaches to the study of leadership and consider a last perspective: one based on social constructionism (or constructivism: both forms are in common use). All the research we have mentioned so far is based on an observer's view of leadership. The participants' own notions about leadership in general or about who is a leader or a follower are regarded as of secondary importance, solicited only to gather data about leadership roles. For example, as we have seen, much of the research following the style approach rested on data gathered from followers about their leaders. These data were then used to identify the characteristics of leadership behaviour. Whether there was any consensus among the followers about what constituted leadership, or even who the leader was, was not considered an interesting question. Once the data had been collected, it was aggregated over all respondents, thus ignoring any differences between them.

From the perspective of these approaches, this may be a reasonable way of proceeding. Social constructionists, however, argue that what is important is participants' own perceptions and accounts. The argument goes like this.[3] (Potter 1996) Concepts like 'leadership' do not exist in a vacuum; they are developed and applied for specific purposes. For social scientists, that purpose is to find out something about group dynamics, but this activity is parasitic on the notions of leadership held by members of society. Unless chief executives, managers and football teams were concerned about leadership, it is unlikely that social scientists would be much bothered with, or perhaps even possess the term. The important issue in this view is therefore to explore how leadership is used in ordinary discourse.

There is a tendency, especially in ordinary conversation, to think of a 'leader' as somehow existing outside and prior to someone's description. That is, a description (for example, "Jane is our leader") is the result of matching up some 'out there' characteristics of Jane's behaviour, attributes or whatever with the linguistic term, 'leader'. Alternatively, however, we could see leadership as being constructed in the course of the speaker producing an account, a report offered in a situation where the identity of the leader has some significance to the participants. Statements such as the (hypothetical) "Jane is our leader" are always designed for some context. Perhaps an external authority is asking about who is taking responsibility for the group, or perhaps group members are trying to decide who is do some job. In either of these and in many other contexts, the statement "Jane is our leader" not only 'describes' the world, but also 'constitutes' it. The statement itself is an action that does a job (such as pointing

[3] For an excellent and carefully argued introduction to social constructionism, see Potter (1996).

out the person who should be taking responsibility and at the same time disclaiming one's own responsibility). It is in this sense that constructionists speak of accounts. In addition, accounts need to be warrantable. If John alleges that Jane is the leader, John needs to be prepared to provide his reasons for saying so. If adequate reasons are not forthcoming, all sorts of inferences may be made about John, that he was trying to evade his own responsibilities, that he was trying to do Jane a favour and so on.

It is important to be clear that this is not a cognitivist perspective. It is not suggested that people have some kind of inner cognitive representation of 'leadership', which is then applied at appropriate moments in interaction. As Conte and Castelfranchi put it:

> Members of the social systems are [not] *observers* who categorize and interpret the (social) world, rather [they are] *agents* who explore it in order to take advantage of or adjust themselves to it. (emphasis in original) (Conte and Castelfranchi 1995, 5)

Since descriptions, such as the assertion that a particular person is a leader, are used by agents to perform actions in the world, such descriptions are motivated to achieve some particular goal and are rhetorical, that is, they are produced in the course of some discourse in order to argue for, demonstrate or support a particular point. The non-cognitivist, indeed anti-cognitivist, stance of this approach follows from two claims. The first is that interaction is jointly and socially constructed and that focusing on individual cognition is therefore to work at the wrong level of analysis. The second claim is that all theories about cognitive representations (such as those involving the identification of intentions and beliefs or the following of scripts and schemas) trade in an unreflexive way on the concepts which ordinary people use in interaction, such as thoughts, feelings, attitudes and so on. In doing so, they elevate these commonsense categories to some special status. Social constructionists are just as interested in the everyday construction of these cognitive descriptions as they are in descriptions of action and they treat them in the same way. (Potter 1996, 103-104)

What follows from this point of view? First, it puts a different complexion on the studies we have previously mentioned. Instead of assuming that there actually are people who are unproblematically leaders and trying to find the definitive set of characteristics by which they can be discovered, leadership is seen as a category which can be useful in discourse. The description of someone as a leader is considered as a context-specific action provided by some agent in order to perform some act. Second, it follows that at different times, in different contexts, the same speaker may provide quite different characterisations of their leader and the various members of a group may describe different persons as their leader. The question of attribution consistency becomes an empirical matter, rather than an unquestioned assumption. Third, the stress on description as action implies that we need to move away from the examination of leadership in the abstract. Instead, the topic of interest shifts to investigations of how and why particular categorisations are made. How does a group member construct an acceptable description that Jane is the leader? What are the implications of providing such an account?

While this is a radical move, it is not as radical as some have suggested. It is not the case, for example, that the social constructionist perspective is solipsistic. It is not asserted that everyone has their own different concept of leadership, nor that anyone can apply the category 'leader' to anyone else without constraint. As a member of western, English speaking culture, we all know what is meant by the word leader, and if there is any doubt about it, we can always refer to the dictionary. On the other hand, the dictionary only gives guidance: I can still use the term metaphorically. Furthermore, while I am free to apply the concept of leader to anyone (or anything), any such application, because it is (part of) an action, allows others to draw inferences which may be to my advantage or to my disadvantage. If I wish, I can assert that Britney Spears is the President of the United States, but it is only because I offer this statement in the context of an example in an academic paper that I can get away with it without you thinking that I am ignorant, confused, silly or mad. So another avenue of research is to look at how particular categorisations are contested and defended. What are the social implications of calling someone a leader? For example, it may be the case that if one member of a group asserts that Jane is the leader, another group member will find it rhetorically easier to go along with this suggestion in consequence, and harder to make a different proposal about the identity of the leader 'stick'. Indeed, we may find that the leadership of long-lived groups ultimately becomes institutionalised, with routines for confirming and recording the leader's identity, so that a major upset is required to change the consensus which has developed.

A model of leadership from the social constructionist perspective will lead us far from most existing social simulation work and so the following description can only begin to suggest some of the issues which will need to be considered. Firstly, as we have emphasised, we are considering members' own concepts of leadership. In the previous simulations we have discussed, there was no explicit representation of leadership available to the agents themselves. Even in the goal-transformation model, the agents themselves did not plan with respect to leaders and followers, but with respect to behaviour that would maximise their sugar intake. Some operators may have required them to act as though they were leaders or followers, but this categorisation of their actions is one made by the observer, not by the agents themselves.

Secondly, the ascription of leadership roles by members (agents) is an action, performed for some purpose and designed for a context. The simulation must include some plausible reason for an agent to engage in this ascription. To be interesting, the simulation needs to be concerned with the agents' use of the categories of leader and follower, not with the consequences of group formation *per se*. Let us therefore modify the simulation framework and award additional sugar to those agents about whom there is a consensus view that they are indeed 'leaders'. In this simulation, then, leaders are not only made but also crowned. Since we are looking for consensus, we are also required to design some kind of communication protocol so that agents can influence each other and eventually come to agree on who is their leader.

Thirdly, we assume that there is a pre-existing generic concept of leadership which agents are already provided with and do not have to develop themselves. This is the equivalent of the dictionary definition. However, like a dictionary definition, the agents' concept of leadership does not, indeed, cannot be so complete that it determines unambiguously its own application to concrete instances. We could model the definition of leadership as a frame with slots to be given values when it is instantiated in a particular context. The slots would be filled by each agent according to its own 'perceptions' of such matters as the number and direction of interactions between agents, inferences about which agents are following which others, the past history of the agents, and the views of other agents about who is the leader. Each slot in the leader frame would be derived by the firing of inference rules which would draw upon the values of other slots and 'observations' of the agent's environment. These rules could be organised within a standard production system with a working memory and a inference conflict resolution system.

Some of the inferences would need to be concerned not just with making 'factual' deductions, but also with the moral consequences of particular conclusions. By 'moral consequences', we mean the inferences which other agents might draw about the agent's motives for ascribing leadership to a particular agent. For example, if agent A proposes that agent B is the leader, but at the end of a process of negotiation, the consensus is that agent C is the leader, inferences could be made about the 'trustworthiness' or 'intelligence' of agent B. If agent A proposes itself as the leader, an inference could be made about the self-serving nature of this proposal, unless other agents had also come to the conclusion that agent A should be the leader and so on. Note that the notion of action is bound up closely with responsibility and morality. It is therefore fitting that a simulation which treats discourse, and in particular categorisation, as action should deal explicitly with moral inferences.

With this in mind we can now remove the initial rather artificial idea that leaders (as defined by the consensus) should be rewarded exogeneously by additional sugar. Instead, we can substitute the top-level goal that agents seek to maximise their own moral standing. Provided that leaders manage to obtain additional approval from other agents by virtue of being leaders, we need no further motivation for leadership in the model.

2. Conclusion

Clearly this description of a simulation to model a social constructionist approach to leadership is very lightly sketched and there are many significant details which still need to be filled in. However, enough has been said to demonstrate the essential paradox that lies behind it. The paradox is that the further we go, the more apparent it becomes that what we are doing is constructing a simulation that purports to be a model of real social processes. When one comes to evaluate the model, the natural criterion is to ask whether group processes really work in the way suggested. This is paradoxical because the model was developed from a position, social constructivism, which explicitly rejects realism as an epistemology. Realism assumes that there are real mechanisms operating in the social world, indeed that there is a 'social world' out there to model, while constructivism is to some extent necessarily relativist (Burningham and Cooper 1996, Sarbin and Kitsuse 1994). In building a social constructionist simulation, we seem obliged to abandon the methodological basis of constructionism and jump straight back into realism. It is not clear how this can be avoided while still sticking to simulation as a research strategy.

In this chapter, we have briefly considered three epistemological positions: positivism, realism and conventionalism, and suggested that the changing currents of leadership research can be related to them. One purpose of this exploration of leadership studies (and their underpinning epistemologies) was to show how simulation could be used to illuminate all of the major approaches. Simulation can open up interesting questions, whether one is positivist, realist of social constructionist in orientation. However, we conclude that there is nevertheless a paradox in this suggestion, for it seems that a research strategy in which simulation features pushes us inexorably towards realism, regardless of our starting point. We conclude that it is indeed unavoidable that research involving social simulation adopts a realist epistemology.

References

Bryman, A. (1996). Leadership in organisations. In Clegg, S., Hardy, C. and Nuro, W. (eds.) *Handbook of Organisational Studies*, Sage, London, pp. 276-291.

Burningham, K. and Cooper G. (1996). Misconstructing constructionism: a defence of social constructionist approaches to environmental problems. Conference on *Social theory and the environment*, ISA RC/24, Woudschoten, The Netherlands, March 20-22, 1997.

Conte, R. and Castelfranchi, C. (1995). *Cognitive and social action*. UCL Press, London. p.144.

Crutchfield, J.P. (1992). Knowledge and meaning: chaos and complexity. In L. Lam and V. Naroditsky (eds.) *Modelling complex phenomena*. Springer-Verlag, New York.

Doran, J.E. (1997). Sensitivity analysis: exploring the parameter space of an artificial society, Dagstuhl Seminar on Social Science Microsimulation, May 5-9 1997, Schloß Dagstuhl, Germany.

Doran, J.E. (2000). Trajectories to complexity in artificial societies: rationality, belief and emotions. In Kohler, T.A. and Gummerman, G.J. (eds.) *Dynamics in Human and Primate Societies.* Oxford University Press, New York.

Epstein, J.M. and Axtell, R. (1996). *Growing artificial societies: social science from the bottom up*. Brookings Institution Press/The M.I.T. Press, Cambridge, MA.

Fiedler, F.E. (1967). *A theory of leadership effectiveness*. McGraw-Hill, New York.

Fikes, R.E. and Nilsson, N.J. (1971). STRIPS: a new approach to the application of theorem proving to problem solving. *Artificial Intelligence*, 2, pp. 189-208.

Halfpenny, P. (1997). Situating simulation in sociology, *Sociological Research Online*, 2(3), http://www.socresonline.org.uk/socresonline/2/3/9.html.

Olson, M. (1965). *The logic of collective action*. Harvard University Press, Cambridge, MA.

Opp, K.-D., Voss, P. and Gern C. (1996). *Origins of a spontaneous revolution: East Germany, 1989*. University of Michigan Press, Ann Arbor, Mi.

Potter, J. (1996). *Representing reality: discourse rhetoric and social construction*. Sage, London.

Sarbin, T.R and Kitsuse J.I. (eds.) (1994). *Constructing the social* Sage, London.

Scott, J. (1995). *Sociological theory: contemporary debates*. Edward Elgar, Cheltenham. p. 173.

Stoghill, R.M. (1950). Leadership, membership and organization. *Psychological Bulletin*, 47, pp. 1-14, quoted in Bryman, *op. cit.*, p. 276.

Stoppard, T. (1968). *Rosencrantz and Guildenstern are Dead*. Faber and Faber, London. pp. 14-15.

SOCIAL INFLUENCE AMONG AGENTS

The Simulation of Social Psychological Theories

HANS-JOACHIM MOSLER
Abteilung Sozialpsychologie
Universität Zürich
Zürich, Switzerland

and

WERNHER BRUCKS
Abteilung Sozialpsychologie
Universität Zürich
Zürich, Switzerland

Abstract

Social influence is an elementary aspect of human societies. People influence others and are influenced by friends, work colleagues, neighbours and even chance acquaintances. In this way, personal attitudes, the attitudes of groups and the attitudes of larger aggregates (like societies) are formed. This makes social influence a constitutive element of societies and one that must be taken into account in simulations. In the field of social psychology, there are several theories that specify the variables and processes of human interaction. However, these theories have found little representation in the literature on simulation. In response to this deficit, we discuss simulations that make social influence a central topic using examples to illustrate the procedures and results. One of our own projects, a simulation of the theory of social comparison explicates the challenges and findings of simulations grounded in social psychology. This serves to demonstrate that the simulation of social psychological theories can be a fertile approach to theory development and can enrich multi-agent simulations.

1. Introduction

In the social sciences, the simulation method must meet the challenge of contributing to the resolution of problems in the real world. For example, we may be interested in finding out why people in a large city like Zürich do not use public transportation more and tend to use their own cars despite the stress and delays involved. In investigating such an issue, implementing the traffic system of a city dynamically on the computer is nothing new. We can also model the social system (the drivers and their behaviour) on the computer.

N.J. Saam and B. Schmidt (eds.)
Cooperative Agents. Applications in the Social Sciences, 125-147

Under favourable conditions, when both the technical and social systems are modelled adequately, the simulation will reproduce the traffic patterns of the city. While the model will correctly represent the real behaviour of the socio-technical system, it is hardly able to pinpoint the causes of this behaviour or find solutions to the corresponding problems. The reason is that the model is valid in its behaviour, but not necessarily in its structure. The question of validity is a very important (but often neglected) subject in social simulation (van Dijkum et al. 1999, Stanislaw 1986). If simulation is going to be used to find the causes of a real world phenomenon then the structure and rules of the phenomenon must be modelled as simply as possible but also as precisely as necessary. This challenge becomes even more pressing if concrete interventions are to be derived from the results of the simulation. Not every social simulation even considers the demand that the structure of the real system should be reproduced in a more or less valid way. If it does, however, it should be tested for validity in the following four areas:

- How valid is reproduction of the structure of the physical environment surrounding the agent? This demand for validity can sometimes be fulfilled comparatively easily if the physical system is well understood. (This is the case for an aeroplane for example.) It is considerably more difficult, however, to reproduce a natural environment with complex (and in part unknown) characteristics. This is the case, for example, with the use of a natural resource where exogenous dynamics play an important role (Smithson 1999).
- How valid is the reproduction of the inner structure of the agent? To do justice to this issue, theoretical knowledge from various disciplines (mainly psychology and sociology) can be of great help. A number of studies have already been published which use theory derived modelling of agents (Latané 1996, Jager, Janssen and Vlek 1999, Deadman, Schlager and Gimblett 2000).
- How valid is the reproduction of interaction between an agent and its physical environment? In this area, some approaches have also been developed that contain, for example, agents' perceptions (Urban 2000) or the formation of mental maps (Janssen and Jager 1999).
- How valid is the reproduction of interaction between one agent and another? Most simulations do not address this last issue adequately. For this reason, it forms the core of the present paper.

The interaction among agents is, besides the other validity issues mentioned above, the most important characteristic of an agent-based computer simulation. The unique strength of agent-based simulation is generally seen to be the fact that social dynamics (such as the rise and fall of cultures) can be modelled from the bottom up and thus explained *post hoc*. (For examples, see Gilbert and Troitzsch 1999, Dean et al. 1999, Marney and Tarbert 2000 and Epstein and Axtell 1996). Agent-based simulations claim to reveal (and sometimes to explain) how it is that on the basis of simple regularities among agents, novel (and at times surprising) collective patterns of behaviour can emerge. It is this that makes agent-based simulations an excellent heuristic tool for generating new knowledge. This knowledge can be implemented in concrete measures provided the validity criteria mentioned above are fulfilled. Often, however, application to practice is not only not attempted, but it is not even the ultimate goal. The present contribution will examine the last of the four points listed above as test criteria for validity. We believe that it is in the modelling of agent-agent interaction that the greatest deficit in empirical know-how is found. In social simulations, the influence of agents upon one other has seldom been modelled on the basis of more or less plausible assumptions. We will examine examples of such plausible assumptions below. Social psychology, however, has at its disposal well founded theories that can explain various forms of interaction. These include the Elaboration Likelihood Model (Petty and Cacioppo 1986) and the Heuristic Systematic Model (Chaiken 1980). There are also theories of social comparison (Festinger 1954) and of normative and informational influences (Deutsch and Gerard 1955). As an alternative to a procedure based on abstractions, it is worthwhile to implement such theories as a concrete simulation on the computer. This simulation becomes available as a module that can be built into other simulations in which agent-agent interaction plays an important role. An example of a workable simulation of psychological theories of social interaction will be presented here.

In the following sections, we define how social influence can be understood among agents in a simulation. Using examples, we will show how much of this understanding has (and has not) been realised in social simulations. The core of this contribution is a presentation of our theory based approach to the modelling of social influence, which is illustrated using a model of group processes. Finally, we draw some conclusions about the advantages and disadvantages of the approach.

2. Implementing social influence in computer simulation

To understand the way that existing social simulations commonly model social influence, it is first necessary to define the phenomenon of social influence. For this, we can turn to a well-known description offered by Latané. Social impact is:

> "any of the great variety of changes in physiological states and subjective feelings, motives and emotions, cognitions and beliefs, values and behaviour, that occur in an individual, human or animal, as a result of the real, implied, or imagined presence or actions of another individual" (Latané 1981, 1996).

What does this mean with regard to the simulation of social influence? Even this simple definition reveals that people exchange information in these processes. The definition makes it clear that social influence is a tangible physical process that does not necessarily occur directly, but under certain conditions can also occur via a medium. For a computer simulation, this means that two conditions must be fulfilled if we are to speak of social influence:

- The agents in a simulation must be capable of accessing information through other agents. In simulations, agents do not normally give information to other agents. This runs counter to our intuitions. The reason lies in programming techniques. Instead, one agent "picks up" information, as it were, from another agent. If information cannot be exchanged then social influence among agents is not possible. A further significant aspect is who can get information from whom. In small social systems, connections among the agents are usually quite simple. For example, all agents may be in contact with each other. Social connections in large populations, however, can pose a difficult challenge to modelling. In some cases these connections must be realised in the form of a complex social network (Mosler and Tobias 2000).
- In addition to information exchange, the information received (which is normally represented as the value of a variable) must be processed according to definite rules in a way that results in a change in agent behaviour. If the model of an agent takes the effects of social influence into account, then the model must include exogenous variables. These are variables that come from the outside, namely from another agent, that the agent must process internally. An agent's actions, attitudes or beliefs are examples of these influences. *Figure 1* shows a systematic representation of this understanding of social influence in an agent-based simulation.

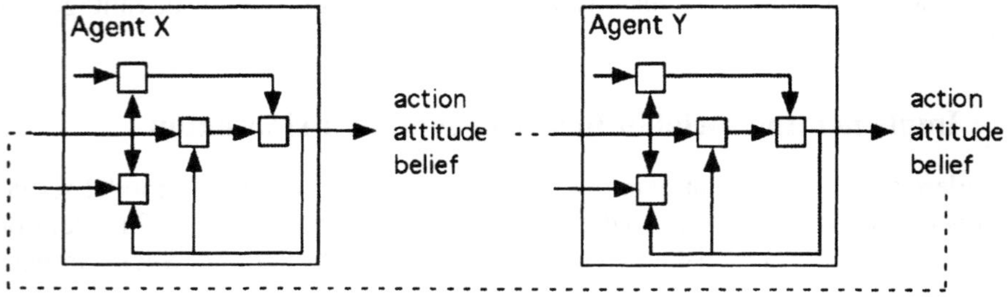

Figure 1: A simulated social system with two agents.
Each agent takes in information (for example, an action, attitude or belief) from the other and transmits to the other corresponding information. Thus, a process of mutual social influence takes place.

In many of the social simulations we are familiar with, social influence takes place in some form or other, even if the information exchanged is very simple in nature. In an example of an early simulation, the classic segregation model of

Schelling (1971), this information consisted of a single, dichotomous feature: black or white cells. In most cases, simulations make simple assumptions about the way in which the transmitted information affects the recipient. In the Schelling model, the assumption was that agents move out of their neighbourhoods when the ratio of black and white cells does not accord with their preferences.

Other examples of assumption guided approaches to the phenomenon of social influence are simulations where the recipient reacts by processing the social information according to game theory rules. In pairs, agents play a dilemma game (such as the Prisoner's Dilemma), determine their advantages and thus their actions for the next round of the iterative game (Liebrand and Messick 1996). Here there is a distinction between an evaluation phase (in which the information received is processed) and an action phase (in which the action chosen during the evaluation phase is performed). This type of simulation makes an important contribution to our understanding of the effects of individual behaviour on collective phenomena. What such simulations do not explain (since they were not designed for it) is how social influence functions *per se*. In other words, how information in one agent can lead to changes in another agent. As we have already mentioned, some assumptions are usually made without any empirical or theoretical basis. However, social psychology provides a number of theories that specify the relevant variables in human interaction as well as the processes that accompany it.

Even if the inner workings of agents are good models of real individuals, the group phenomena that emerge at the macro-level will hardly be valid where agent-to-agent interaction has not been implemented on a solid theoretical basis. If simulation aims to model and explain behaviour in a real population, then a theory of interaction among the agents is indispensable. Despite the many social simulations that neglect this point, some have begun to tackle this problem. We will now describe two representatives of this new approach, where interaction among agents takes place according to psychological theories.

Janssen and Jager (1999) take an integrated approach to simulation by equipping their agents with behaviour rules that are based exclusively upon well-founded theoretical knowledge in psychology. In one of their simulations, agents (known as "consumats" or artificial consumers) have to choose between two different products competing on the market. The extent to which the social surroundings of the agent influence its actions is determined by well known psychological mechanisms. Consumats have four fundamental behaviours at their disposal. Firstly, considered action (choice theory or theory of reasoned action) which comes into play when the current need and the likelihood that it can be satisfied are both great. Secondly, social comparison (social comparison theory and theories on relative deprivation) will be carried out when the current need is great and there is little guarantee that it can be satisfied. Thirdly, behaviour will simply be repeated (classical and operant conditioning theory) when the need is small and the guarantee of satisfaction is great. Finally, the behaviour of other agents will be imitated (social learning theory and theory of normative conduct) when both the need and the guarantee of satisfaction are small. Using these rules, the simulation defines whether social influence takes place and what its effects are on the basis of well-known and accepted theories. Yet this is still not the implementation (as a simulation) of an actual theory of social influence such as social psychology can provide.

That step has been made by Latané (1996), who has developed an agent-based simulation using his own theory of "social impact". This led to the development of a theory of "dynamic social impact" (Latané 1996, Nowak and Latané 1994). Represented as cellular automata, 400 agents differ in a number of dichotomous attitudes like their political orientation (right versus left). The agents have various degrees of influence that they exert mutually according to Latané's theory. The theory indicates that the total influence (I) acting upon an agent is the multiplicative product of the strength of the influence of the other agents (S), their spatial and temporal immediacy (I), and the simple number of agents supplying influence (N). The resulting simple formula is I = f(SIN) and this is used as the rule for continuing interaction by the agents in the simulation. The influence working on every single agent is calculated for each of the two values of the dichotomous variables. The agent accedes to the stronger influence. It may, for example, change its own political orientation from the left to the right if the social influence of right wing political opinion is stronger than that of left wing opinion. If not, the agent will retain its existing orientation. Nowak and Latané demonstrate that under most conditions, an agent population in which social influence operates according to their theory will achieve a state of balance. Minorities do survive and the population does not take on a uniform attitude. This line of research is (to our knowledge) the only attempt to implement a known psychological theory of social influence in a computer simulation with the exception of our own simulation work in Zürich. One of our simulations will be presented in more detail below as a representative example of our overall approach.

Both the examples above take a step in the right direction. In each population, the social influence operating on the agents is modelled on the basis of recognised psychological theory. In the case of the consumats, the theories only form a background for the formulation of the actual rules governing social interaction. By contrast, Nowak and Latané realise a direct computer implementation of a theory of social influence.

The following discussion will introduce our own approach to the modelling of social influence. We begin with a framework model that represents behaviour towards the environment and specify the factors and processes involved. This model was designed with the aim of suggesting how the largest number of people in a population can be moved to think and act in a more environmentally friendly way. This framework model clearly shows the social psychological theories which were implemented. From these theories, we will take one example in order to explain the modelling and simulation of social influence in group processes. Although the framework model focuses on the modelling of social influence, it also represents interaction with the physical world as well as with the wider social environment.

2.1. THE FRAMEWORK MODEL

We designed a basic model of an individual. The simulation involves 10,000 identically structured copies of this individual equipped with different individual characteristics. This population serves as the basis for the simulated influence and resource use processes (see *figure 2*).

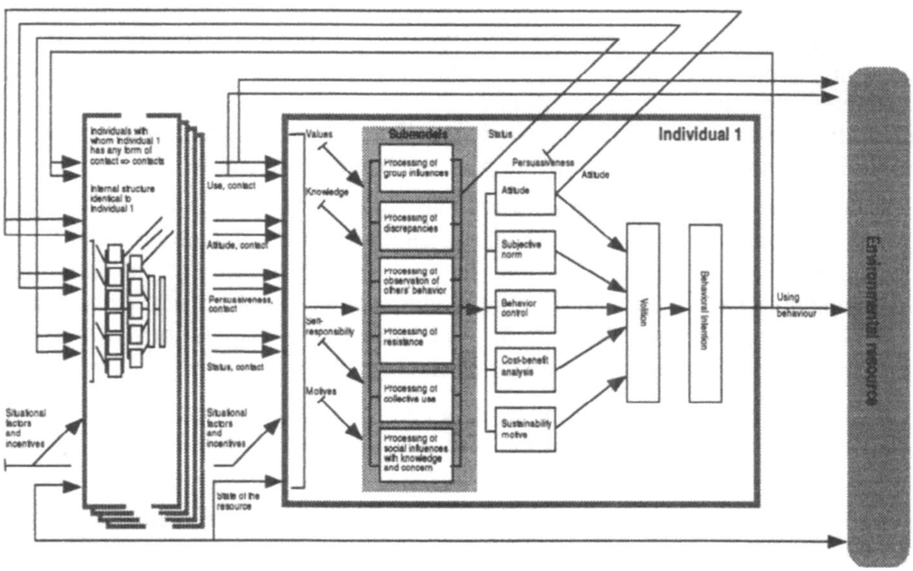

Figure 2: Framework model of behaviour towards the environment

The agents differ only in the values of their attributes. They all function according to the same social psychological principles. These principles are based on a few central and well-founded theories. The framework model begins by specifying the input and output variables of theory based processing by individuals. The model yields information about the inner psychological processes that take place when people use environmental resources or influence each other mutually, whether deliberately or accidentally. Inner psychological processes are triggered as people communicate with one another in daily life and observe themselves and others.

These processes change depending on further internal and external conditions, the ways in which we feel, think and argue about the environment and the way we act towards it. The simulated individuals have differently structured networks of social contact: the number of friends, acquaintances, neighbours and strangers they observe.

The following section describes the external and internal variables as well as the inner processes of the agents.

2.2. EXTERNAL INPUT VARIABLES

External input variables are the influences that are exerted on a person from the outside and are perceived by that person in some form. Possible distortions in the perceptions of individuals are not taken into consideration in this model. For each individual, some output variables function as external input variables for other individuals with whom the individual has 'contacts' (see *figure* 2). In addition, some non social input variables, such as specific situational parameters and the momentary state of the resource being used, enter into the calculation of an individual's output value. The following description presents the variables and processes shown in *figure* 2 from left to right:

- **Use, contact**: A summary of all environmentally related actions performed by a contact. These actions towards the environment are conceived as resource using actions. For resource use, a particular resource can be entered into the model with its characteristic parameters. Resources include water, air, and wild game populations while the corresponding parameter might be the rate of regeneration of a resource.
- **Attitude, contact**: A summary of a contact's opinions and evaluations of environmental issues, objects and actions towards the environment as expressed in various ways.
- **Persuasiveness, contact**: The intensity and quality with which a contact makes a case for various attitudes related to the environment. While a sign is given for an attitude ("for" or "against" environmental protection) persuasiveness indicates the intensity and quality with which these attitudes are presented.
- **Status, contact**: This variable represents the sum of a contact's relevant personal resources. (These include social competence, trust, prestige, knowledge, power, possessions and so on.)
- **Situational factors and incentives**: These include all influences from the societal-institutional surroundings (rules and prohibitions, positive and negative incentives and the like). From the entire spectrum of possible actions, situational factors produce a subset of actions.
- **State of resource**: The state of the environment or of a particular environmental resource, as "noticed" and established by the person.

2.3. INTERNAL INPUT VARIABLES

Internal input variables have an effect on simulated psychological processes "from the inside". The emergence of individual values for these variables (as a result of the individual's learning history) is not examined in this simulation.

- **Values**: Stable orientation with regard to environmental facts, objects and actions.
- **Knowledge**: The extent and quality of information about the environment such as knowledge of the regeneration parameters for a specific resource.
- **Self-responsibility**: Describes the extent to which people attribute responsibility to themselves for the state of the environment as opposed to holding other persons, organisations or institutions responsible.
- **Motives**: A summary of various motives, such as curiosity and laziness, that enter into a person's readiness to act in certain ways. Depending upon these motives, an action - independent of its consequences - will be viewed as "desirable" or "easy".

3. Processing of Input Variables in the Sub Models

External and internal input variables are processed in different sub models according to the theory being applied. The output values of the sub models may have a direct outward effect on other persons with whom the individual has contact (attitude, persuasiveness). Alternatively, they may affect processes leading up to behaviour and decisions on the use of resources and consequent actions. The following sub models were designed and simulated (see also *figure 2*):

I. Processing of group influences upon attitudes towards the environment. This is an application of the theory of social comparison processes. (Festinger 1954, Frey, Dauenheimer and Parge, and Haisch 1993.)

II. Processing of discrepancies between environmental behaviour and environmental attitude. This is an application of dissonance theory (Festinger 1957, Frey and Gaska 1993.)

III. Processing of observations of others' behaviour towards the environment. This is an application of Bandura's social learning theory (Bandura 1979, 1986 and Stalder 1985).

IV. Processing of resistance to environmental protection measures. This is an application of reactance theory (Dickenberger, Gniech and Grabitz 1993, Wicklund 1974, and Brehm and Brehm 1981).

V. Processing of information on the collective use of an environmental resource. This is an application of commons dilemma research. (Compare the summary of findings in Dawes 1980, Diekmann 1991, Liebrand, Messick and Wilke 1992, and Messick and Brewer 1983.)

VI. Processing of communicative influence in view of the individual's feeling of concerned consternation about the state of the environment, knowledge about the environment and biases. This is an application of the Elaboration Likelihood Model (Petty and Cacioppo 1986, Stahlberg and Frey 1993, Mosler, Schwarz , Ammann and Gutscher in press.)

3.1. PROCESSES LEADING TO BEHAVIOUR

Processes leading to behaviour are direct preliminary steps towards behaviour. Theories of resource mobilisation (Klandermans 1984) and planned behaviour (Frey, Stahlberg and Gollwitzer 1993, Ajzen 1991, and Ajzen and Madden 1986) indicate five components playing important roles in behavioural intentions towards the environment:

- Attitude towards the environment as an attitude towards environmental protection and individual behaviour regarding the environment
- The subjective norm as a "perception of pressure from the social surroundings". This expresses the expectations that confront the individual and the degree to which the individual is willing to fulfil these expectations
- Behavioural control: The subjective conviction that one is in fact able to perform the behaviour
- Cost-benefit analysis as a motive for behaviours, which results from weighing the direct costs and benefits of actions
- Sustainability as a motive, leading to willingness to restrict personal use of a resource. This readiness depends upon both the absolute value that a person places on an environmental resource and on the current discrepancy between collective sustainable use and the actual pattern of use shown by others. The higher the subjective value of the environment or the smaller the discrepancy between actual and sustainable patterns of utilisation, the greater the effect of the sustainability motive in affecting a person's own restraint in use.

These components all influence behavioural intention, which leads to behaviour towards the environment. In addition, we introduce volition into the model, which is closer to actual behaviour than behavioural intention. Psychological research (Frey et al. 1993) has shown that the act of making a decision commits a person to the "when" and "where" of attempting to realise an intention.

In what follows, we will take a detailed look at sub model I (dealing with social comparison processes) which we used to model the social influences operating in groups.

4. The processing of group influences upon attitudes towards the environment: application of the theory of social comparison processes

In everyday life, group discussions and decisions play a central role. Collective events influence the thinking and actions of individuals, the group and even the entire social system. The individual, as part of the group, codetermines the events within the group, which develops its own dynamics. These dynamics in turn determine the thinking and behaviour of group members. The aggregation of group thought and action towards the environment, as well as the interaction among groups, determines what happens in the social system. This in turn produces patterns of use for environmental resources. Groups are linked together by individuals with multiple group membership or those who have fleeting contacts between groups.

Frey, Dauenheimer, Parge and Haisch (1993, 114ff) have developed an integrative concept of social comparison processes. This concept encompasses Festinger's theory of social comparison and Tajfel's theory of social identity. The final component is the notion of comparison levels for alternatives proposed by Thibaut and Kelley.

According to this conception of social comparison (ibid. 111ff), people in groups change their attitudes depending on the existing pressure to conform, the attractiveness of the group and the perceived threat to their self concept if they change their position. Attitude change results from the pressure to conform weighted by the attractiveness of the group. If the self-concept is threatened, however, people will not change their attitudes. Threats to a person's self-concept arise when his or her values differ to a certain degree from the average attitudes of other members of the group. Individuals perceive too great a discrepancy between their own values and the group's attitude as threatening. In order to adapt to the group, the individual would have to deviate too far from their personal values. Under social pressure, they would become untrue to themselves.

Figure 3 shows the model. People are more likely to change an attitude the more attractive the group is and the greater the group pressure to conform. Attitudes will change in the direction of the average attitude of the group. Attitude change occurs, however, only if the change does not pose a threat to the self-concept. The attractiveness of the group results from two components. The first is the difference between a person's attitude and the average attitude of the group: a group is more attractive if the person and the group are on the same "wave length". The second component is the average status of the group: its prestige, power and social resources.

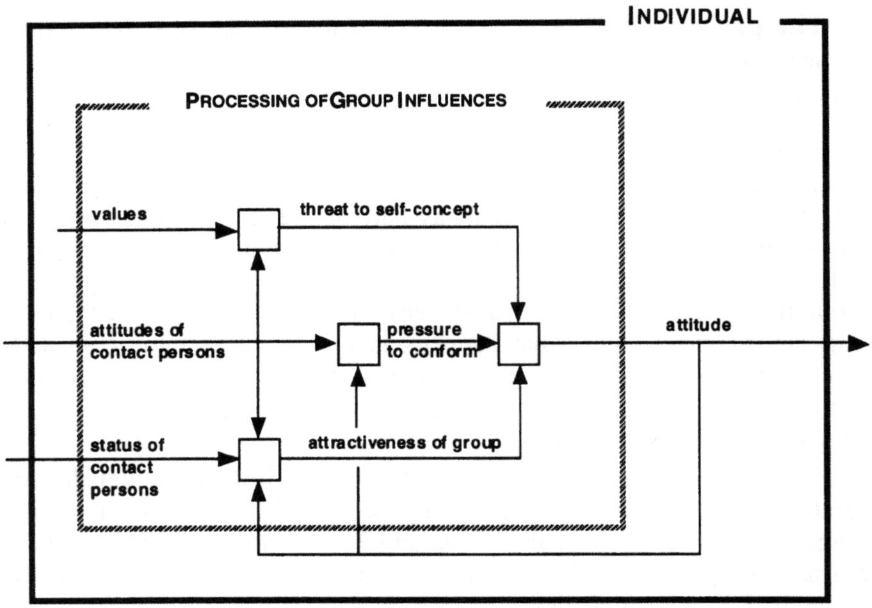

Figure 3: Representation of the implemented processes in the sub model "processing of group influences". Arrows show the variables and processes are represented by the boxes.

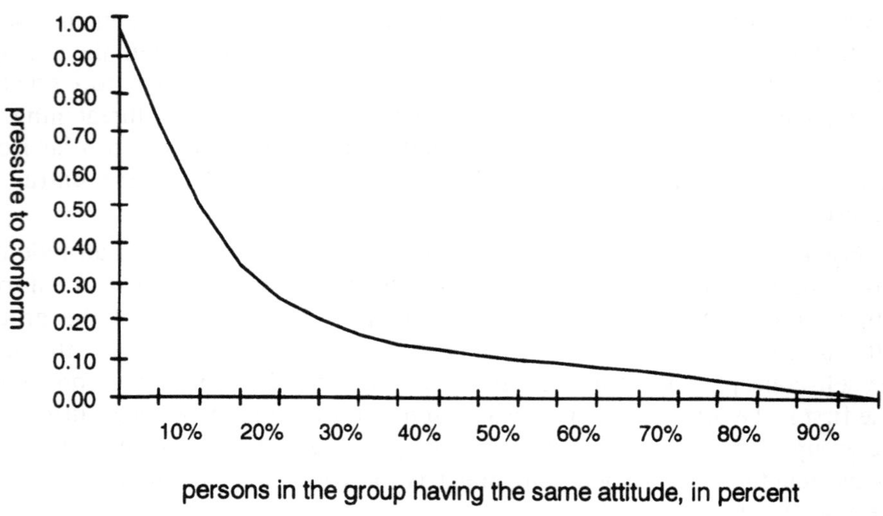

Figure 4: Pressure on the individual to conform as a function of group attitude sharing. If the individual shares an attitude with less than 50% of the group, the individual is a member of a minority. The majority attitude is held by over 50% of the group.

The conformity pressure of a group can be calculated according to Tanford and Penrod's (1984) well documented formula. Pressure to conform, expressed in terms of the total number of group members, is calculated as a separate exponential function for minority and majority members (see *figure 4*).

4.1. SIMULATION EXPERIMENTS WITH SMALL GROUPS

To develop a better understanding of social influence on groups, we present simulation experiments with small groups of three and ten people. It is important to bear in mind that this sub model (like the theories it is based on) was developed with a strong group focus. Accordingly, individuals always compare their attitudes to the *average* attitude of the other members of the group, in the form of a sub group average and not to the attitude of other individual members of the group. Furthermore, the individuals always change their attitude in the direction of the sub group average. For the individual group member, the other members form the reference group or social standard. In every round of the simulation, the group influences each of its individual members.

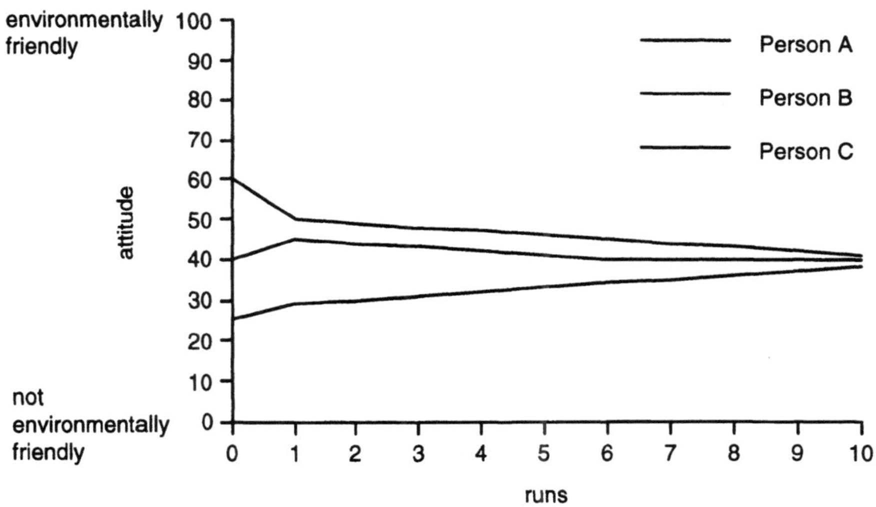

Figure 5: Conformity.
A starts with an attitude of 60, B an attitude of 25 and C of 40.
The value of average group attitude falls from 42 to 40.

All variables range from 0 to 100. For variables related to environmental orientation, a score of 100 indicates maximum environmental friendliness and 0 indicates maximum lack of environmental friendliness. A score of 50 represents neutrality. For all simulations, the values of an individual have the same value as their attitudes. The self-concept is threatened if the difference between a person's values and the average group attitude is greater than 30.

In the group simulations, the threat to the self-concept proved to be a crucial factor. If there is no threat to an individual's self-concept, all individuals will converge on the group average (see *figure 5*).

If there is a threat to the self-concepts of a minority, the majority adapts to the attitude of the minority (see *figure 6*). Here we find that A does not change attitude for quite a while. This is because s/he experiences a threat to the self-concept as long as the average attitude of the rest of the group differs from his or her value (of 75) by 30 or more points. A's adherence to his or her own attitude position has the effect of moving the group's average attitude from an initial 30 to 60.

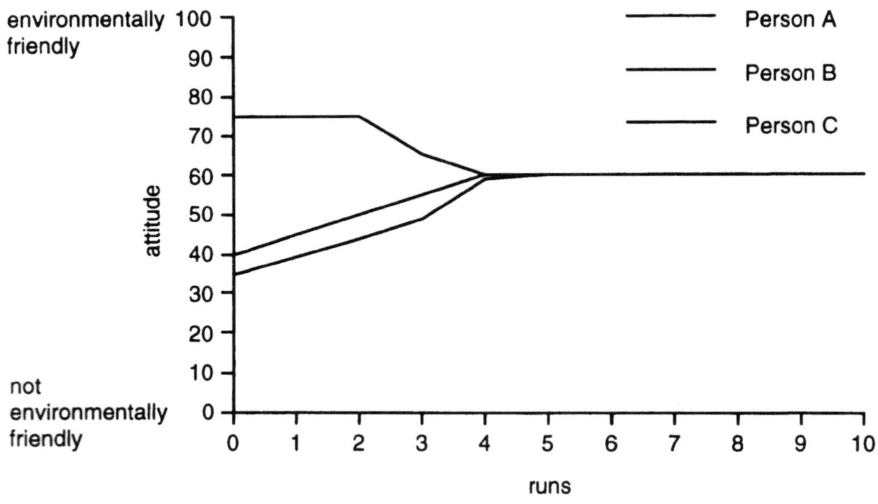

Figure 6: Minority influence.
A starts with an attitude of 75, B with 35, and C with 40.
The average attitude of the group rises from 50 to 60.

In simulation experiments with ten person groups, the proportions in the minority and majority were varied between 1:9 and 5:5. This meant that there were different pressures on group members to conform. A look at the majority's adaptation during the initial iterations in *figure 7* reveals that larger minorities have a stronger influence. If the minority makes up 50% of the group (which makes it barely a minority), the majority changes its attitude faster than it would if the minority made up 40% of the group and so on.

Astonishingly, it is the smallest minority that has the strongest influence on the final attitude of the group (*figure 7*). This can be explained as follows. The larger the minority, the less threatening the entire group is to minority members, because the average attitude of the sub group approaches the values of minority members. As a consequence, the minority tends to adapt its attitude more the greater the minority (compare *figure 8*). This is because such attitude changes do not threaten the self-concept. The smaller the minority, however, the more threatened the self-concept. Minority members' values are so far from the average attitude of the group that they feel threatened in their self-concepts. The majority consequently (after some time) moves towards the minority position until the minority no longer feels threatened and begins to show its own tendency to adapt in the direction of the majority (*figure 8*).

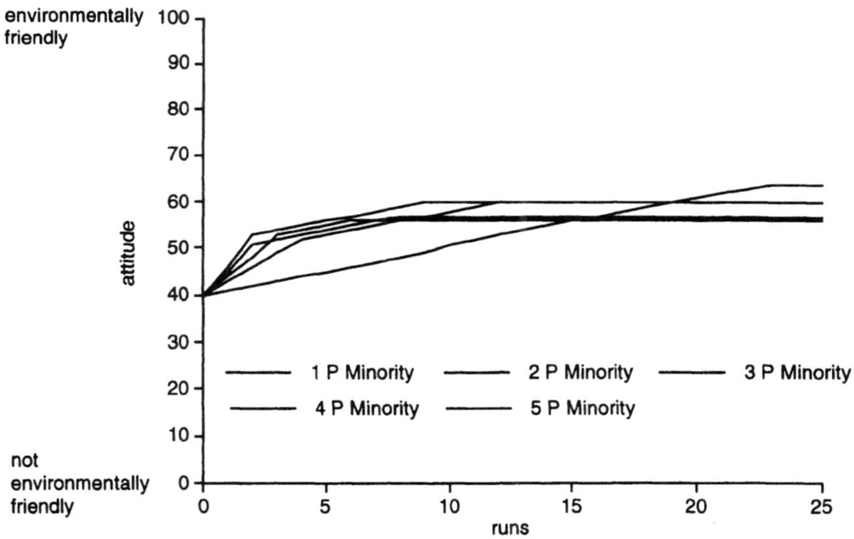

Figure 7: Changes to the average attitude of the minority with varying proportion of minority members (one to five persons) in a ten person group. The initial attitude of all minority members is 80 and the initial attitude of all majority members is 40. Everybody has a status of 40. Simultaneous changes in the majority are shown in *figure 8*.

If there are one to three members in the minority, this adaptation process only begins after the majority members have changed their attitudes in the direction of the minority – that is, when the initial threat to the self-concept of minority members no longer exists. If there are three to four minority members, the adaptation process in minority members begins immediately as they experience no threat to their self-concept. The greatest delay in the adaptation process takes place if the minority makes up only 10% of the group (one person). The least delay occurs if the minority makes up 40% of the group (four persons, compare *figure 8*). The intersections and overlaps of the curves in *figures 7* and *8* are caused by the interplay of the effects of varying conformity pressures with different proportions of majority and minority and the changing effects of threats to the self-concept. Both figures show clearly, however, that the influence of minority and majority groups on each other does not vary in linear fashion with the proportions of group members.

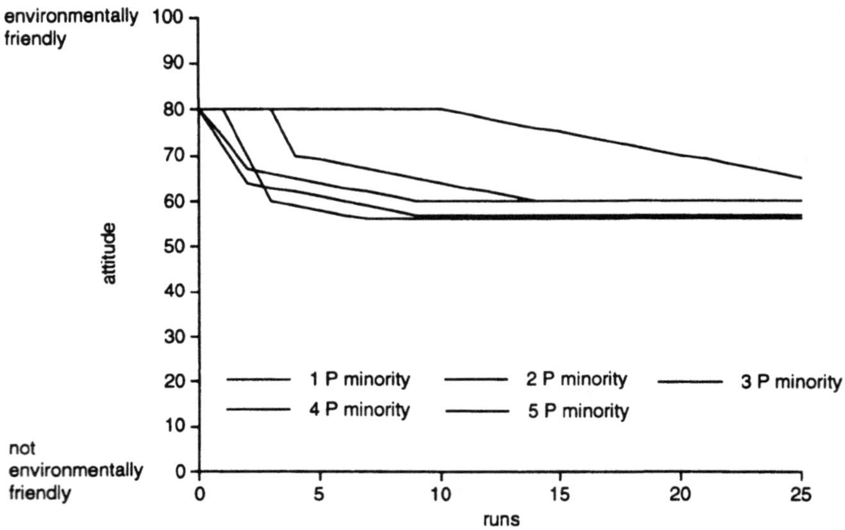

Figure 8: Changes to the average attitude of the minority with varying proportion of minority members (one to five persons) in a ten person group. The initial attitude of all minority members is 80 and the initial attitude of all majority members is 40. Everybody has a status of 40. Simultaneous changes in the majority are shown in *figure 7*.

There have been numerous confirmations of the fact that minorities exercise influence provided they are consistent and unwavering in their behaviour. (See Clark and Maass 1990 and Wood, Lundgreen, Oulette, Busceme and Blackstone 1994.) The threat to the self-concept explains the high consistency of minorities that Maass and others have highlighted as an important factor in influence (Clark and Maass 1990, Moscovici 1980 and Moscovici and Doise 1994). Minorities lose influence, on the other hand, if they are unmoving and rigid in behaviour (Maass, West and Clark 1985 and Nemeth, Swedlund and Kanki 1974). The effects of consistency and rigidity can also be seen in our simulations and can be explained through the varying threat to self-esteem. The minority remains consistent in its attitude until the attitude of the majority has come so close to its own that there is no longer a threat to the self-concept. At this point, the minority adapts to the majority. This shows a certain willingness to adapt (flexibility) and is therefore not seen as rigid. If the minority is unmoving, however, and maintains its attitude unwaveringly, the majority stops adapting to the minority position because it would have to move too far from its own original position and because it would then experience a threat to its own self-concept.

4.2. SIMULATION EXPERIMENTS WITH POPULATIONS

Let us turn to experiments with populations of 10,000 persons. The research question for these simulation experiments is as follows. How must a minority of people behaving in an environmentally friendly way be distributed and networked within a population in order that the environmentally unfriendly majority comes to change its attitudes and behaviour?

The population is structured into groups of 10 persons, such that each group is in contact with two other groups. We assume that an environmental campaign can make 10% of the population embrace more friendly attitudes towards the environment than previously (an increase of 25 points). This occurs for a limited period of time. (For example, for one week which is one step in the simulation). These persons we call (attitude) "pioneers". Pioneers are also more strongly sensitised to deviations from their own positions. Because of this, they have a particularly low susceptibility to attempts at social influence. If other individuals in the group make even minor deviations from their own (pioneer) positions, this is taken as a threat to the self-concept of the pioneer.

As an additional experiment, pioneers are organised in different ways. Following the short environmental campaign:

- they remain in their groups (isolated) and have no wider contacts with those outside
- they remain in their groups but are in contact with other pioneers (networks)
- they form their own groups (core groups) which may have few or many contacts with other groups

In the graphs presenting the findings, average behaviour towards the environment (or the average attitude towards the environment by the population) is always shown on the ordinate. If a curve rises, this means that the population is shifting in the direction of environmental responsibility. On the abscissa we show the simulation steps. A step involves a complete calculation cycle. All individuals exert influence on their contacts or are influenced themselves and – together with the others – they then utilise a resource.

For the experiments, we usually began with populations that were rather environmentally unfriendly in attitude (having a mean of 40) with the values of the variable for individuals distributed around this mean.

In order to escape from the criticism that we have built the results of the simulation into the program, all experiments were conducted according to a baseline/control group design. The results of a number of steps with and without manipulations (control group) form the basis for comparison and populations were always identical (which is only possible in computer simulation). In the following section, some important issues in social intervention and campaign planning are examined with the aid of the simulation model.

Figure 9 shows the experiments and their results. The intervention applies only to the fifth step. The dynamics described below result from this short-term intervention.

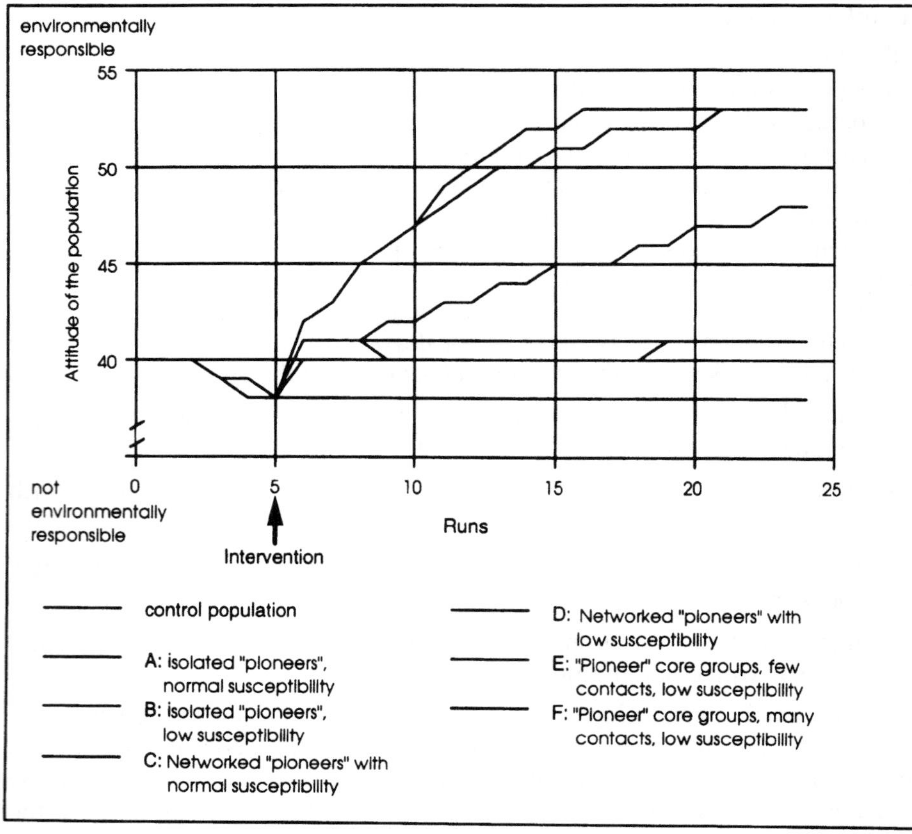

Figure 9: The change in average attitude for the population according to various group oriented forms of intervention as described in the text.

A. In each group there is an isolated pioneer whose susceptibility to influence is the same in degree as that of the members of the group majority.

B. In each group there is an isolated pioneer whose susceptibility is lower than that of the members of the group majority.

C. In each group there is a pioneer whose susceptibility is the same in degree as that of group members. In addition, the pioneer is in contact with 10 other pioneers in other groups. Thus pioneers no longer perceive themselves to be a minority.

D. As in case C, but here the pioneers have a lower degree of susceptibility.

E. All pioneers are concentrated within their own core groups and show low susceptibility to influence. Each pioneer is in contact with another person outside the core group.

F. As in case E except that pioneer group members are in contact with 10 other persons outside of the core group.

5. Results

a) It is extraordinarily important for pioneers' to have low susceptibility to influence. This prevents them adapting too soon to the surrounding majority.

b) If their susceptibility is not low, there will be no positive effects from being networked with other pioneers and thus perceiving themselves as no longer in a minority.

c) Core groups (also with low susceptibility) only have an effect on the surrounding majorities if they have numerous contacts outside the core groups. (Curve F in *figure 2* continues to climb in later steps but not as high as curves B and D.)

In summary, pioneers have to hold more environmentally responsible attitudes and their susceptibility to influence must be low. In effect, they must be quick to perceive threats to their self-concept. Pioneers must be activated or active in as many groups as possible and should not be concentrated in just a few groups. Further measures are not required.

6. Discussion

What has been gained by basing a model of social influence on social psychological theories? We believe that the following points deserve attention:

1. Theory guided modelling is closer to reality because the theories are based on real findings.
2. A simulation based on theories is easier to test for validity, because the findings that the theory is based on can be used for this purpose.
3. The findings from theory based simulations can be fed back into the

research on a particular theory. Researchers working in a particular field are then in a better position to evaluate the validity and heuristic value of the simulation.

4. Theory based simulation also contributes to the further development of a theory because of the necessity for formalisation. Specifications produced in the simulation framework (as well as extensions to theories) can then be tested using traditional methods, experimentally for example.
5. Theories should integrate many findings which means that tried and tested theories are rich in implications. As a result, theory based simulations can also reveal the most diverse implications.
6. Through theory guided knowledge and modelling of the "inner" workings in the individual, social influence processes can be understood better. For real world problems, attempts to change the behaviour of groups or populations can be improved.

We also recognise a number of disadvantages in theory based simulations of social influence:

1. Each social psychological theory has any number of components and further developments that cannot be taken into account by a given simulation in a complete and differentiated manner. As a result, it is a difficult undertaking to crystallise the core statements from a theory.
2. The social psychological theories to be used show a high degree of specification and are thus difficult to understand for people who do not work in the field. To model these theories, therefore, you have to be in the field yourself or work with someone who is.

All in all, the advantages of modelling social influence processes on the basis of social psychological theories strongly outweigh the disadvantages. We appeal to those interested in simulation not to overlook this fund of theories about social influence. Making full use of this material will bring great gains to simulation as well as to theoretical research.

References

Ajzen, I. (1991). The theory of planned behavior. *Organizational Behavior and Human Decision Processes,* 50, 179-211.

Ajzen, I. and Madden, T.J. (1986). Prediction of goal directed behavior: attitudes, intentions, and perceived behavioral control. *Journal of Experimental Social Psychology*, 22, 453-474.

Bandura, A. (1979). *Sozial-kognitive Lerntheorie.* Klett-Cotta, Stuttgart.

Bandura, A. (1986). *Social foundations of thought and action.* Prentice-Hall, New York.

Brehm, S.S. and Brehm, J.W. (1981). *Psychological reactance: A theory of freedom and control.* Academic Press, New York.

Chaiken, S. (1980). Heuristic versus systematic information processing and the use of source message cues in persuasion. *Journal of Personality and Social Psychology,* 39, 752-766.

Clark, R.D. and Maass, A. (1990). The effects of majority size on minority influence. *European Journal of Social Psychology,* 20, 99-117.

Dawes, R.M. (1980). Social dilemmas. *Annual Review of Psychology,* 31, 169-193.

Deadman, P.J., Schlager, E. and Gimblett, R. (2000). Simulating Common Pool Resource Management - Experiments with Adaptive Agents Employing Alternate Communication Routines. *JASSS - Journal of Artificial Societies and Social Simulation,* 3(2), <http://jasss.soc.surrey.ac.uk/JASSS/3/2/2.html>.

Dean, J.S., Gumerman, G.J., Epstein, J.M., Axtell, R., Swedlund, A.C., Parker, M.T. and McCaroll, S. (1999). Understanding Anasazi Culture Change Through Agent-Based Modeling. In: Kohler, T.A. and Gumerman, G.J. (eds.). *Dynamics of Human and Primate Societies: Agent-Based Modeling of Social and Spatial Processes*. Oxford University Press, New York.

Deutsch, M., and Gerard, H.B. (1955). A study of normative and informational social influences upon individual judgement. *Journal of Abnormal and Social Psychology*, 51, 629-636.

Dickenberger, D., Gniech, G. and Grabitz, H.-J. (1993). Die Theorie der psychologischen Reaktanz. In: Frey, D. and Irle, M. (eds.), *Theorien der Sozialpsychologie.* Band I: Kognitive Theorien. Hans Huber, Bern, 243-274.

Diekmann, A. (1991). Soziale Dilemmata Modelle, Typisierungen und empirische Resultate. In: Esser, H. and Troitzsch, K.G. (eds.). *Modellierung sozialer Prozesse.* Informationszentrum Sozialwissenschaften, Bonn.

Epstein, J.M., and Axtell, R. (1996). *Growing Artificial Societies: Social Science from the Bottom Up*. MIT Press, London.

Festinger, L. (1954). A theory of social comparison processes. *Human Relations*, 7, 117-140.

Festinger, L. (1957). *A theory of cognitive dissonance.* Stanford University Press, Stanford, CA.

Frey, D., Dauenheimer, D., Parge, O. and Haisch, J. (1993). Die Theorie sozialer Vergleichsprozesse. In: Frey, D. and Irle, M. (eds.), *Kognitive Theorien* (2). Hans Huber, Bern, 81-122.

Frey, D. and Gaska, A. (1993). Die Theorie der kognitiven Dissonanz. In: Frey, D. and Irle, M. (eds.), *Kognitive Theorien* (2). Hans Huber, Bern, 275-326.

Frey, D., Stahlberg, D. and Gollwitzer, P.M. (1993). Einstellung und Verhalten: Die Theorie des überlegten Handelns und die Theorie des geplanten Handelns. In: Frey, D. and Irle, M. (eds.), *Kognitive Theorien (2).* Hans Huber, Bern. 361-398.

Gilbert, N. and Troitzsch, K.G. (1999). *Simulation for the Social Scientist.* Open University Press, Buckingham.

Jager, W., Janssen, M. and Vlek, C.A.J. (1999). *Consumats in a commons dilemma: Testing the Behavioral Rules of Simulated Consumers.* COV Report Number 99-01, Centre for Environment and Traffic Psychology, University of Groningen.

Janssen, M. and Jager, W. (1999). An Integrated Approach to Simulating Behavioural Processes: A Case Study of the Lock-In of Consumption Patterns. *JASSS - Journal of Artificial Societies and Social Simulation,* 2(2), <http://jasss.soc.surrey.ac.uk/JASSS/2/2/2.html>.

Klandermans, B. (1984). Mobilization and participation: Social-psychological expansions of resource mobilization theory. *American Sociological Review*, 49, 583-600.

Latané, B. (1981). The psychology of social impact*American Psychologist,* 36, 343-356.

Latané, B. (1996). Dynamic Social Impact. In: Hegselmann, R., Mueller, U., and Troitzsch, K.G. (eds.), *Philosophy and Methodology of the Social Sciences.* Kluwer Academic Publishers, Dordrecht, 287-310.

Liebrand, W.B.G., Messick, D.M. and Wilke, H.A. (eds.) (1992). *Social Dilemmas: Theoretical Issues and Research Findings*. Oxford: Pergamon Press.

Liebrand, W.B.G., Messick, D M. (1996). Computer Simulations of Sustainable Cooperation in Social Dilemmas. In: Hegselmann, R., Mueller, U. and Troitzsch, K.G. (eds.), *Modelling and Simulation in the Social Sciences from the Philosophy of Science Point of View.* Kluwer Academic Publishers, Dordrecht.

Maass, A., West, S.G. and Clark, R.D. (1985). Soziale Einflüsse von Minoritäten in Gruppen. In: Frey, D. and Irle, M. (eds.), *Theorien der Sozialpsychologie, Band II: Gruppen- und Lerntheorien.* Hans Huber, Bern, 65-91.

Marney, J.P., and Tarbert, H.F.E. (2000). Why do simulation? Towards a working epistemology for pratitioners of the dark arts. *Journal of Artificial Societies and Social Simulation*, 3(4), <http://jasss.soc.surrey.ac.uk/JASSS/3/4/4.html>.

Messick, D.M. and Brewer, M.B. (1983). Solving Social Dilemmas: A Review. In: Wheeler, L. and Shaver, P. (eds.), *Review of Personality and Social Psychology.* Sage, Beverly Hills, 11-44.

Moscovici, S. (1980). Towards a theory of conversion behavior. In: Berkowitz, L. (ed.), *Advances in experimental social psychology.* Academic Press, New York, 209-239.

Moscovici, S. and Doise, W. (1994). *Conflict and Consensus: A general theory of collective decisions.* Sage, London.

Mosler, H.-J., Schwarz, K., Ammann, F. and Gutscher, H. (in press). Computer simulation as a method of further developing a theory: Simulating the Elaboration Likelihood Model (ELM). *Personality and Social Psychology Review.*

Mosler, H.-J., and Tobias, R. (2000). Die Organisation kollektiver Aktionen durch Beeinflussung der individuellen Teilnahmeentscheidung. *Kölner Zeitschrift für Soziologie und Sozialpsychologie*, 52, 264-290.

Nemeth, C.J., Swedlund, M. and Kanki, B. (1974). Patterning of the minority's response and their influence on the majority. *European Journal of Social Psychology*, 4, 53-64.

Nowak, A., and Latané, B. (1994). Simulating the Emergence of Social Order from Individual Behaviour. In: Gilbert, N. and Doran, J.E. (eds.), *Simulating Societies: The Computer Simulation of Social Phenomena.* UCL Press, London, 63-84.

Petty, R.E., and Cacioppo, J.T. (1986). The Elaboration Likelihood Model of persuasion. In: L. Berkowitz (ed.), *Advances in Experimental Social Psychology Volume 19*, Academic Press, New York, 123-205.

Schelling, T.C. (1971). Dynamic Models of Segregation. *Journal of Mathematical Sociology*, 1, 143-186.

Smithson, M. (1999). Taking Exogenous Dynamics Seriously in Public Goods and Resource Dilemmas. In: Foddy, M. (ed.), *Resolving Social Dilemmas: Dynamics, Structural and Intergroup Aspects.* Psychology Press, Philadelphia, PA.

Stahlberg, D. and Frey, D. (1993). Das Elaboration-Likelihood-Modell von Petty und Cacioppo. In: Frey, D. and Irle, M. (eds.), *Kognitive Theorien (2).* Hans Huber, Bern, 327-360.

Stalder, J. (1985). Die soziale Lerntheorie von Bandura. In: Frey, D. and Irle, M. (eds.), *Gruppen- und Lerntheorien.* Bern: Hans Huber, 241-272.

Stanislaw, H. (1986). Tests of Computer Simulation Validity. *Simulation and Games*, 17(2), 173-191.

Tanford, S. and Penrod, S. (1984). Social influence model: A formal integration of research on majority and minority influence process. *Psychological Bulletin*, 95(2), 189-225.

Thibaut, J.W. and Kelley, H.H. (1959). *The social psychology of groups.* New York: John Wiley.

Urban, C. (2000). PECS - A Reference Model for the Simulation of Multi-Agent Systems. In: Suleiman, R,. Troitzsch, K.G. and Gilbert, N. (eds.), *Tools and Techniques for Social Science Simulation.* Physica-Verlag, Heidelberg.

van Dijkum, C., de Tombe, D. and van Kuijk, E. (1999). *Validation of Simulation Models.* SISWO, Amsterdam.

Wicklund, R.A. (1974). *Freedom and reactance.* Wiley, New York.

Wood, W., Lundgreen, S., Oulette, J.A., Busceme, S. and Blackstone, T. (1994). Minority Influence: A meta-analytical review of social influence processes. *Psychological Bulletin*, 115 (3), 323-345.

ENROLLING SOFTWARE AGENTS IN HUMAN ORGANISATIONS

The Exploration of Hybrid Organisations within the Socionics Research Program

INGO SCHULZ-SCHAEFFER
Institut für Soziologie
Technische Universität Berlin
Berlin, Germany

Abstract

To a considerable degree, complex organisations are characterised by internal incoherence, by differences between the individual departments' preferences and patterns of actions and by inconsistencies of the organisation's goals. By contrast, information technology is intended to operate in a coherent way. In consequence, information systems (designed to support organisational processes) usually proceed from the assumption of organisational coherence. The more traditional approaches to implementing information systems in organisations face a dilemma. Either they produce dysfunctional restrictions imposed on the organisation by the information system's need to establish firm cause/effect relationships. Or, conversely, they lead to incoherence imposed on the information system by the organisation's structure. With agent technology, however, we have a chance to arrive at organisational information infrastructures that allow for accommodation on both sides. The paper describes the theoretical background of the INKA research project (which is part of the socionics research program funded by the DFG). The project aims to develop solutions to the problem sketched above by modelling and exploring hybrid forms of interaction between humans and artificial agents within organisational settings.

1. Introduction

Multi-agent systems are distinguished from most previous technological devices and approaches by one basic feature. The components they consist of, the agents, can be designed to interact in a way that imitates mechanisms of social interaction between human actors. This specific feature of multi-agent systems has to be considered alongside an observation that applies to technology in general. It is a well-known fact from social studies of technology that technological innovation necessarily comprises the development of a new socio-technical system. Each new technology has to meet certain conditions imposed by the social context (interests, practices or social rules that influence the user's

N.J. Saam and B. Schmidt (eds.)
Cooperative Agents. Applications in the Social Sciences, 149-163

behaviour) in order to be of use. Together, these two points lead to an interesting, but so far mainly unexplored question. Do we face a new type of socio-technical system in the field of multi-agent technology? It seems that this system is characterised by the fact that not only the interrelations between the participating humans, but also the interrelations of the technical components are structured (or can be structured) by means of social interaction. Will the walls between social action and technical performance break down as a result, confronting us with socio-technical hybrids, systems that "do not distinguish between interactions with humans and interactions with artificial agents" (Parunak 1996: 150)?

There are good reasons (which I will turn to later in section 6) why it is appropriate to start with social interaction in organisations, if one wants to explore hybrid forms of interrelations between humans and agents. To this end, the INKA project (**In**tegration **k**ooperationsfähiger **A**genten in komplexen Organisationen)[1], a research project within the new socionics research program, has been established. In the present paper, I want to describe the general idea of this project (section 3) and show that organisational software agents are interesting objects of research. This is true not only for computer scientists looking for innovative technological solutions (section 7) but also for organisational researchers concerned with the question of how to adjust organisational structures to information technology or vice versa (section 5). Last but not least, it is of interest for social scientists in the field of technology studies, who are investigating the interrelations between technology and society (section 4). But let me begin with a short outline of the socionics research program.

2. Socionics as frame of reference

Socionics aims to model and explore artificial societies. These are social settings that cannot be established or understood without considering the actions of artificial agents (cf. Malsch et al. 1996; Müller et al. 1998). It is a new interdisciplinary field of research that combines sociological knowledge and multi-agent technology in three different, but interrelated ways:

Firstly, socionics is concerned with the question of how to use mechanisms of coordination and social exchange between human actors to develop innovative strategies of multi-agent problem solving. Here, the rationale is the same as it is in bionics. Take a principle you have found to structure biological events (or in our case social events), look to see if there is a technological problem where this principle may help to provide a solution and transform it

[1] The members of the project team are: Hans-Dieter Burkhard, Werner Rammert, Frank Janning, Gabriela Lindemann von Trzebiatowski, Ines Münch, Klaus Scheuermann and the author.

into a technological procedure. Human societies can be viewed as a vast reservoir of methods of joint problem solving between distributed entities (human actors). If you know their structuring principles, i.e. if you have sociological concepts describing them adequately, then there are excellent opportunities to use them as models of joint problem solving among distributed technical entities (artificial agents).

In the first area of socionic research, sociological knowledge serves as a means of developing multi-agent systems. In the second strand of socionics the means-end relation is reversed. Here, the aim is to employ multi-agent systems as tools for the simulation of human society or parts thereof, and to use these simulations to improve sociological concepts and theories. The basic assumption is that multi-agent systems are "more able to 'mirror' societies and groups of people than their alternatives" (Doran and Gilbert 1994: 10). One of the main advantages of multi-agent simulation is in dealing with the micro-macro gap, one of the central problems in social theory. Multi-agent simulation allows us "to cross the micro-macro bridge" (Drogoul and Ferber 1994: 6), that is to model interrelations between individual action and social structure. This opens up new opportunities for testing sociological hypotheses and may help to overcome some of the restrictions of previous work in the field of social simulation.

The third research perspective combines aspects of the first two fields of study and adds a new focus of interest directed at hybrid systems of human actors and artificial agents. I have already raised the main questions of this strand of socionics in the first paragraph of this paper. How do socio-technical systems change, if they consist of humans and agents that interrelate their behaviour by means of social interaction? What are the soci(ologic)al and technological preconditions and consequences, the prospects and risks of such a development?

The idea that it might be fruitful for both sociology and multi-agent research to bring their fields of study together was discussed in the eighties by a group of American computer scientists and sociologists (cf. Strübing 1998). But it took almost another decade to invent a name for the then "Unnamable" field (Bendifallah et al. 1988). The term "socionics" was coined by Thomas Malsch (cf. Malsch et al. 1996). He and the author developed the framework of socionics outlined above. This led to the socionics research program funded by the Deutsche Forschungsgemeinschaft (DFG-Schwerpunktprogramm Sozionik). Within this research program (which started in the autumn of 1999) there are eight projects, the INKA project being one of them. Each project consists of an interdisciplinary team of sociologists and computer scientists. Each one has its main focus on one of the socionic perspectives discussed above. In the case of our project, it is the third perspective that provides the focus and I am a member of the sociological sub team.

3. The INKA project

Generally speaking, the INKA project aims to explore the conceptual and technological foundations for agent-based systems which are able to deal successfully with problems of the incoherence and heterogeneity of complex organisations. The aim is not only to think about useful multi-agent architectures, but about hybrid structures: forms of interaction that are characterised by interrelations between the activities of human actors and artificial agents. A research project like ours can easily lead to mere speculation. To avoid this, it is crucial to base the issues under consideration on empirical research. Therefore, we chose to focus on a concrete domain of application: the field of medical therapy planning within the complex organisation of hospitals.

To a considerable degree complex organisations are characterised by internal incoherence, by differences between the individual departments' preferences and patterns of action and by inconsistencies in the goals of the organisation as a whole. By contrast, information technology is intended to operate in a coherent way. For this reason, information systems (which are designed to support organisational processes) usually proceed from the assumption of organisational coherence. The discrepancy between the actual incoherence of complex organisations and the coherence which information technology both assumes and requires leads to a dilemma. Either organisational structure will be forced to adapt to the structure of the information system or vice versa. In the first case, the consequence is that social mechanisms for dealing successfully with organisational incoherence (such as negotiation, coalition formation, the strategic use of information and the existence of informal structures) will be overruled or invalidated. This weakens the organisation's capabilities for self-adjustment because it would be naïve to believe that information systems could cause organisational incoherence to disappear. In the second case, the information system will adopt incoherent features of the organisation (such as inconsistent knowledge) and that will affect its performance.

We hope to contribute to the solution of this problem by reconceptualising computer supported organisations as hybrid systems of human actors and artificial agents. With agent technology there is an opportunity to process organisational data and support organisational decision-making without the need to homogenise the relevant data and the information infrastructure first. The strategy is to replace the globally planned information system with sets of software agents. Each agent, in its interactions with other agents, represents the interests of its own user (an individual or collective actor). In this way each agent can be designed as a coherent piece of software, that is to say as a representation of consistent sets of interests and patterns of actions held by an organisational unit or individual. On the global level, however, the missing coherence will at least partially be compensated for by the agents' capabilities for interaction.

The design of agents as human assistants in contexts of social interaction necessarily has to deal with two conflicting goals. On one hand, the agents are supposed to do what they are intended to do. On the other, they should be capable of behaving adequately in their respective interactions with their users. Hence, the designers have to face a trade-off between restricting the agents' programs of action and conceding them a certain amount of autonomous decision-making ability. The same trade-off also applies to the structuring of human action in organisations. Taking this similarity as a basis for our research, we hope to contribute to the development of an empirical foundation for (as yet) mainly hypothetical questions about the advantages and limitations of hybrid cooperation.

4. The agency of technology in socio-technical hybrids

Usually, we assume technology to be a more or less useful means, a tool controlled by human users which is well constructed if it produces the expected effects in a predictable and reproducible manner. From this point of view, the claim that technological devices may possess agency seems to be an absurdity or science fiction at best. Nevertheless, on closer inspection, the everyday perspective turns out to be too narrow even in the face of well established technology because it does not take into account interdependencies in the behaviour of technological devices and human actors.

Charles Perrow took an important step in the direction of a new perspective on technology in his work on organisations. He not only considers human actors, their routines and programs of action to be the components of complex organisations, but technical components – pumps, mains or computers – as well. Perrow holds that in complex organisations (power plants, chemical factories and hospitals) the overall behaviour of the system is defined more by characteristics of relations between heterogeneous components than by the actions of a single human or machine (Perrow 1986: 146 ff.). By dissolving the traditional boundary between human and material components of organisations, he is able to successfully explain structural causes of organisational accidents. This is a persuasive argument for his conception of technology.

Several researchers in the field of science and technology studies have taken further steps in this direction. With his historical reconstruction of the development of the electric power system, Thomas Hughes has exemplified the advantages of a view that focuses on interrelationships between heterogeneous components of socio-technical systems (Hughes 1987). But the most radical approach in this line of argument is the actor network theory developed by Michel Callon and Bruno Latour (Callon 1987; Latour 1988). Here, the distinction between technology and society is completely dismissed. All components of socio-technical ensembles are viewed as "actants", that is as "entities that do things" (Latour 1988: 303). The authors claim that only with such a symmetrical approach is it possible to describe "techno-social imbroglios" (ibid. 309) in an unbiased way. Only by taking this step can we adequately recognise the contributions of non-human actants (technical components) to the development and stabilisation of socio-technical networks (cf. Schulz-Schaeffer 2000).

The actor network theory employs the weakest possible concept of agency. It considers each event that is caused by an identifiable entity and that influences a given state of affairs as its action. In everyday life and in sociology alike, we usually presuppose the satisfaction of at least one additional condition before conceding agency: that the actor could have done otherwise. In the words of Anthony Giddens (1984: 9): "Agency concerns events of which an individual is the perpetrator, in the sense that the individual could, at any phase in a given sequence of conduct, have acted differently." Finally, in human interaction a third criterion can and will be employed if there are doubts concerning questions of agency: "... it is expected by competent agents of others ... that actors will usually be able to explain most of what they do, if asked." (ibid. 6)

The technological devices of everyday life which constitute the main examples used by actor network theory (door closers or seatbelts) are obviously actors only in the first sense. In the case of program controlled technology, the second criterion begins to be applicable, even though their degree of freedom to choose is normally limited to pre-planned alternative modes of conduct. This situation definitely changes with agent technology. The autonomy of the agents to decide between different options and their proactive nature (cf. Wooldridge and Jennings 1995: 2) are taken to be the main point of research and development. In this case, as the BDI approach has shown, it no longer seems so absurd to develop and understand technology by using human-centred concepts such as belief, desire and intention (cf. Haddadi and Sundermeyer 1996).

The agency of technology becomes an even more interesting (and difficult) issue if you take into account the fact there is no single notion of human agency. Rather, each of the three criteria mentioned above represents a partial aspect of the notion. In particular, it is important to understand that not all behaviour has to meet the three criteria in order to pass for an individual action. Even the first concept of action, where agency is nothing more than the actor's ability to make a difference, cannot simply be dismissed as an undue simplification of human affairs. On the contrary, a considerable part of what is usually considered to be human action does not fulfil any conditions other than that the individual in question is regarded as having caused it.

When considering whether or nor it makes sense to attribute agency to a specific entity in a socio-technical ensemble, these are reasons why we should not simply react affirmatively (in the case of humans) or negatively (in the case of technology). Instead, it is useful to assume a continuum of more or less agency on both sides (cf. Rammert 1998: 115f.). Thus agent technology – as the most ambitious attempt to enable technological devices to act – becomes the paradigmatic field for research concerned with technological agency and with exploring hybrid forms of interrelations between human actors and technological devices (cf. Schulz-Schaeffer 1998).

5. Technology in the context of organisational incoherence

Rational choice theories describe organisations as institutions pursuing goals which are well defined by stable and internally consistent sets of preferences. It is assumed that organisational decision-making is always driven by the search for the best way to reach the organisation's goals (cf. March and Simon 1958: 136ff, March 1990: 282f). Empirical studies of organisations, however, support the opposite view that the preferences of organisations are anything but consistent and stable (cf. March 1990: 13f). Some organisations are "characterised by problematic preferences, unclear technology, and fluid participation" (Cohen et al. 1972: 1) to a degree that makes it feasible to view them as "organised anarchies" (ibid.). In contrast to what rational choice theories suggest, such organisations "can be described better as a loose collection of ideas than as a coherent structure" (ibid.). They "operate on the basis of simple trial-and-error procedures, the residue of learning from the accidents of past experience, and pragmatic inventions of necessity" (ibid.). The authors claim these properties of organised anarchy to be particularly conspicuous in public organisations, but nevertheless to be "characteristic of any organisation in part - part of the time" (ibid.). Therefore, a "theory of organised anarchy will describe a portion of almost any organisation's activities, but will not describe all of them" (ibid.).

Organisational incoherence constitutes a particular problem for implementing technology. The problem is that technology relies on managing causal relationships, pinpointing goals and controlling initial conditions in a way that has proven to be unrealistic for a wide range of organisational conduct. Hence, the crucial question is how to ensure the proper functioning of technology in complex organisations, where preferences are often only vaguely defined, different and possibly conflicting patterns of action are employed and shifting coalitions of interest have to be recognised.

According to James D. Thompson, the embedding of technology in complex organisations relies on a specific form of internal differentiation. "Organisations cope with uncertainty by creating certain parts specifically to deal with it, specialising other parts in operating under conditions of certainty or near certainty" (Thompson 1967: 13). Every formal organisation "contains a sub organisation whose 'problems' are focused around effective performance of the technical function ... such as the materials which must be processed and the kinds of cooperation of different people required to get the job done effectively" (ibid. 10). To ensure the efficiency of the technical core, exogenous variables (that may interfere with the technology as a closed system of cause/effect relationships) have to be excluded as far as possible. "Under norms of rationality, organisations seek to seal off their core technologies from environmental influences" (ibid. 19). An example of this strategy is the stockpiling of materials and supplies as a means to buffer input disposal uncertainties (cf. ibid. 20).

However, with the spread of information technology, organisations are more and more confronted with the employment of a kind of technology towards which the strategy of buffering no longer applies. Where information technology is used to support organisational operation, to structure the flow of information and to organise the organisation's memory, technology and organisation often interpenetrate in a way that doesn't allow for a separate technical core. "Aspects, previously held to be distinct: 'technology' (equipment) and 'organisation' (control structure) are increasingly merging ... The so-called new technologies extend technology to organisational processes in such a manner that the sociological notion of technology has to be reconceptualised radically. The organisational and social implications of computer-based information technology in production and in the office are calling for a new sociological definition of technology as 'fait social'." (Behr et al. 1991: 159ff).

There seem to be only two overall strategies for implementing information systems if one has to assume that (in the case of information technology) organisational change and technological change are inseparably connected. The first strategy has been called "systematische Rationalisierung" (cf. Altmann et al. 1986). It is the attempt to use the information technology's power of rationalisation to restructure the organisation's procedures in a systematic way. For that purpose, it is necessary to formalise processes of organisational decision-making accordingly, to integrate heterogeneous patterns of conduct, to align distinct domains of action and so on. The second way involves adjusting the information technology to the given organisation, in other words reproducing the organisational structure as the structure of the information system. Where it occurs, this way of implementing information technology is usually something that just happens unintentionally and incidentally, rather than constituting an explicit strategy.

Both ways of doing things entail some serious problems. The first approach underestimates the features of incoherence mentioned above which are, especially in complex organisations, a continuous source of resistance against attempts to systemise processes and states of affairs (cf. Behr et al. 1991: 26, 46). The most important disadvantage of the second approach is that the incoherence of the organisation's structure is inherited by the information system, but at the same time the technology does not inherit the mechanisms to deal with it. The individual and collective actors in an organisation can cope with inconsistencies and uncertainties of goals, interests and programs of action by using informal modes of coordination, by temporarily forming coalitions and so on. Incoherently structured information systems, on the other hand, only weaken the cause/effect relationships upon which technology is based and lead to a loss of efficiency for the system in question.

6. Organisational software agents as starting points for the exploration of hybrid systems

The development of agents which are capable of acting in a certain way entails a crucial difficulty. On one hand, software agents are supposed to differ from more convenient technological devices in that they are not intended simply to execute predetermined technological procedures, but to perform specific tasks more or less autonomously. On the other hand, these agents are not supposed to possess unbounded autonomy. They shall "not really behave freely as humans may" (Burkhard 1995: 299) but only support the human activity of problem solving. Hence, for a wide range of agent technology applications it seems reasonable to steer a middle course that avoids undue control of agents but undue autonomy as well (Schulz-Schaeffer and Lührs 1998: 22ff).

As stated above, this trade-off between autonomy and control applies in a similar way to the organisation of concerted human action. For this reason organisations are especially well suited to serve as a test case for endeavours to integrate artificial agents into human contexts of action. The overall strategy of the INKA project is to exploit this similarity: to restrict or to open up the artificial agents' capabilities to act in a way which matches the restrictions or degrees of freedom that human actors face in organisations.

Members of complex organisations rarely act as freely as the average competent individual in a modern society is able to. Rather, their agency is restricted to a specific scope of duties, they are urged to follow predefined procedures, the resources at their disposal are limited and the organisation's formal structure regulates what they have to do by order of others. If we understand how the conduct of organisation members is structured by such patterns of restricted action, we can hope to design software agents which fit into the organisation's frame of reference. With respect to the goal of constructing hybrid systems, an important feature of domains with restricted action is that they make it a lot easier to integrate artificial agents. One main problem with developing software agents that are intended to collaborate with humans is to provide them with the capabilities necessary to behave adequately in such a role. Obviously, this problem is less pressing where the social setting in question restricts the involved individuals capabilities to act by itself.

As a rule, domains of restricted action are most visible within the subdivisions of a complex organisation. By contrast, problems caused by inconsistent goals and distinct modes of action occur predominantly where co-ordination *between* departments is involved. Here, especially, we face the kind of organisational co-ordination problems that cannot easily be solved by given rules and procedures. Instead, they demand actors who are somewhat free to evaluate the situation at hand and to act accordingly. Therefore, actors who are confronted with the need to coordinate their department's work with tasks performed by other subunits of the organisation have to develop a double orientation of heteronomy and autonomy. On one hand, they act heteronomously in that they represent their subunits' interests and goals. On the other, they are less committed to the goals of the other departments involved. This opens up a certain space for negotiations. In this way complex organisations are able to establish several technological cores – the domains of restricted action – and at the same time to deal with the problem of internal incoherence.

Agent technology allows us to model this double orientation and thereby to design information technology that works efficiently in the face of organisational incoherence. On one hand, the agents have to be designed to act as representatives of their (individual or collective) users. This means that their capabilities to act are restricted by the represented subunits' rules and procedures of goal attainment. On the other hand, the agents have to be able to represent their users' interests and goals autonomously in the interaction with other (human or artificial) members of the organisation. To inter-relate the components of the information system we do not need a pre-planned overall structure. Instead we can use suitable mechanisms of social exchange and negotiation between agents (and human actors).

The agents' behaviour does not necessarily have to be very complex in order to contribute to an organisation's information infrastructure in such a manner. Where organisations impose extensive restrictions on their members' agency, the entry barriers for integrating artificial agents are comparably low. Therefore, a promising strategy seems to be to begin by modelling hybrid systems in organisational domains where the humans' (and correspondingly the agents') behaviour is regulated to a large extent and then gradually to enhance the agents' capabilities to act.

7. Agent-based medical therapy planning

As a result of internal functional differentiation of hospitals, there are always several physicians, nurses, diagnostic facilities and administrative units involved in the treatment of a patient. Within the different departments plenty of electronic data are generated. However, these data are often filed only in the local databases of different departments. To transmit such data to other departments (or their representatives) they then have to be manually copied, written into paper forms and delivered perhaps by the hospital's pneumatic post system. This is a time consuming procedure and leads to undesirable delays in the patients' treatment. The hospital we chose to be our domain of application employs modern network technology. Nevertheless, incompatibilities between the local databases, the data formats and the query languages used constitute obstacles to electronic transmission of patient related data between the hospital's departments.

It is important to understand that this heterogeneity in the information systems is not simply an engineering problem soluble by means of data integration. Rather, it results to a considerable degree from organisational incoherence in the information infrastructure. Each department has its own routines for dealing with the data generated there, and it is often a matter of habit to prefer a certain database system. Be it for habitual or rational reasons, the kind of data processing employed by a department is consequently believed to be more suitable to the own work processes than the other departments' systems. If it is not possible to eliminate such inconsistencies in practice then we have to look for information technology which supports gathering and exchanging data within a heterogeneous and incoherent information infrastructure and which does not need to integrate that data globally. This is what agent technology promises to be able to do.

The fewest difficulties in assigning tasks to software agents will occur in the case of simple organisational routines. These are tasks that often require a lot of paperwork at present. They include requesting and confirming dates in scheduling a patient's medical treatment, gathering the medical data from different departments, converting heterogeneous data formats into the format used by the department in question and dealing with the problem of updating. (The latter is an important problem if the storage of similar information in different places is unavoidable.) Beginning with these simple tasks we hope gradually to enhance the agents' ability to carry out procedures which require more deliberation such as planning of time schedules, occupancy planning, medication planning and surgery planning.

8. Conclusion

Agent technology offers the promise of a solution for the problem of linking organisation and technology. The more traditional approaches for the implementation of information systems in organisations often lead to a dilemma. Either dysfunctional restriction is imposed on the organisation by the information system's need to establish firm cause and effect relationships. Or, conversely, incoherence is imposed on the information system by the organisation's structure, thus reducing the efficiency of the technology. With agent technology, however, we have a chance to arrive at organisational information infrastructures that allow for accommodation on both sides. In this respect, an important feature of software agents is the new opportunities they provide for connecting an organisation's local and global processes of coordination.

As is shown by a large amount of empirical research on complex organisations, it is not very realistic to treat organisations as rationally structured means designed to achieve well-defined sets of goals. On the other hand, this does not imply that organisations are completely irrational. Instead, we will usually find organisational domains which are governed in an internally consistent way by certain common norms of rationality. These are the domains of restricted action I referred to above. Within such domains, the chances are good that we can establish chains of cause and effect, allowing us to co-ordinate people and processes technically. Outside these domains, chains of cause and effect probably cannot be established and here more complicated forms of coordination will be found.

Unlike more traditional information technology, software agents can be made to fit into this double structure. This involves designing agents which support their local domain's work processes on one hand and, on the other, represent their domain within the organisation's information infrastructure. Within such a setting, the agents are directly assisting only the work of certain actors or groups of actors respectively. This considerably reduces the capabilities they need to perform adequately, because the agents only have to perceive their environment very selectively: only so far as needed to contribute to solving specific tasks within specific domains of restricted action. At the same time, each domain's work is more or less interrelated with work to be done in other parts of the organisation, and this applies to the organisation's information processing as well. The process of co-ordinating the work of different organisational sub units usually takes place under conditions involving heterogeneous local norms of rationality and competing local interests and preferences. Under such circumstances socially negotiated solutions are required. Research on modelling mechanisms of social interaction as means of multi-agent co-ordination offers various ideas which will hopefully turn out to be useful in tackling this kind of co-ordination problem within the information infrastructure of a complex organisation.

As you will have noticed, the preceding considerations do not produce results in the sense of describing a hybrid organisation which has already been modelled. Rather, the purpose of the present paper is to show that there are good reasons not to view hybrid systems as the subject of futuristic speculation, but as a topic of research that has the power to produce both new theoretical insights and to generate innovative practical technologies. In particular, I want to stress three lines of argument:

(1) The traditional notion of technology as a passive means to human ends is no longer sufficient to understand the role of technological devices within socio-technical systems. For this reason, a strand of research has been developed which uses the concept of technological agency to describe the interrelation of technology and society. Agent technology (as an explicit attempt to model agency technically) and hybrid systems (as an attempt to connect human and technological agency) may help to shed light on this debate.

(2) The new capabilities of computer based information technology (which allow them to support administrative and planning processes in organisations) mean that organisation and technology become mingled so that we can no longer strictly separate technical procedures from the surrounding organisational activities. Consequently, the organisation's information infrastructure has to deal simultaneously with two competing requirements: ensuring technical efficiency and handling organisational incoherence. Reconceptualising organisational information systems by implementing software agents seems to be a good way to tackle this problem (and at the same time to explore basic features of hybrid organisations).

(3) It depends heavily on the domain of application whether or not hybrid systems will be developed successfully. On one hand, we must take into account that even the most sophisticated agents' capabilities are far from those we take for granted in the average human actor. Therefore, we have to look for domains of restricted action where there is at least some chance for artificial agents to behave adequately. On the other hand, without looking for tasks where a certain amount of autonomous action is required from the agents in question, there is no reason to use agent technology at all. In both respects, we assume organisations to be the most suitable starting point for integrating artificial agents into contexts of human interaction.

References

Altmann, N., Deiß, M., Döhl, V., and Sauer, D. (1986). Ein „Neuer Rationalisierungstyp" - neue Anforderungen an die Industriesoziologie. In: *Soziale Welt* 37, 191–206.

Behr, M., Heidenreich, M., Schmidt, G., and Graf von Schwerin, H.-A. (1991). *Neue Technologien in der Industrieverwaltung. Optionen veränderten Arbeitskräfteeinsatzes*, Westdeutscher Verlag, Opladen.

Bendifallah, S., Blanchard, F., Cambrosio, A., Fujimura, J., Gasser, L., Gerson, E.M., Henderson, A., Hewitt, C., Scacchi, W., Star, S.L., Suchman, L., and Trigg, R. (1988). *The Unnamable: A White Paper on Socio-Computational 'Systems'*, Research Note 60, DAI Group, University of Southern California, unpublished draft manuscript available from Les Gasser, Department of Computer Science, University of Southern California, Los Angeles.

Burkhard, H.-D. (1995). Agent-Oriented Programming for Open Systems. In: Wooldridge, M.J. and Jennings, N.R. (eds.). *Intelligent Agents*. Proceedings of the ECAI-94 Workshop on Agent Theories, Architectures, and Languages, Springer-Verlag, Berlin. 291-306.

Callon, M. (1987). Society in the Making: The Study of Technology as a Tool for Sociological Analysis. In: Bijker, W.E, Hughes, T.P. and Pinch, T.J. (eds.): *The Social Construction of Technological Systems: New Directions in the Sociology and History of Technology*. The MIT Press, Cambridge, MA, 82-103.

Cohen, M.D., March, J.G., and Olsen, J.P. (1972). A Garbage Can Model of Organizational Choice. In: *Administrative Science Quarterly* 17(1), 1-25.

Doran, J.E., and Gilbert, N. (1994). Simulating Societies: An Introduction. In: Gilbert, N. and Doran, J.E. (eds.): *Simulating Societies: The Computer Simulation of Social Phenomena*, UCL Press, London. 1-18.

Drogoul, A., and Ferber, J. (1994). Multi-Agent Simulation as a Tool for Modeling Societies: Application to Social Differentiation in Ant Colonies. In: Castelfranchi, C. and Werner E. (eds.): *Artificial Social Systems: 4th European Workshop on Modelling Autonomous Agents in a Multi-Agent World*, MAAMAW '92, Springer-Verlag, Berlin, 3-23.

Giddens, A. (1984). *The Constitution of Society: Outline of the Theory of Structuration*. Polity Press, Cambridge.

Haddadi, A., and Sundermeyer, K. (1996). Belief-Desire-Intention Agent Architectures, in: O'Hare, G.M.P and Jennings, N.R. (eds.). *Foundations of Distributed Artificial Intelligence*. John Wiley and Sons, New York, 169-210.

Hughes, T.P. (1987). The Evolution of Large Technological Systems. In: Bijker, W., Hughes, T.P. and Pinch, T. (eds.). *The Social Construction of Technological Systems*. The MIT Press, Cambridge, MA, 51-82.

Latour, B. (1988). Mixing Humans and Nonhumans Together: The Sociology of a Door-Closer. In: *Social Problems* 35(3), 298-310.

Malsch, T., Florian, M., Jonas, M., and Schulz-Schaeffer, I. (1996). Sozionik: Expeditionen ins Grenzgebiet zwischen Soziologie und Künstlicher Intelligenz. In: *Künstliche Intelligenz KI* 2/1996, 6-12.

March, J.G. (1971). Die Technologie der Torheit. In: March, J.G. (ed.) (1990): *Entscheidung und Organisation: Kritische und konstruktive Beiträge, Entwicklungen und Perspektiven*. Gabler, Wiesbaden, 282-295.

March, J.G. (1990). Einführung. Eine Chronik der Überlegungen über Entscheidungsprozesse in Organisationen. In: March, J.G. (ed.): *Entscheidung und Organisation. Kritische und konstruktive Beiträge, Entwicklungen und Perspektiven*. Gabler, Wiesbaden, 1-23.

March, J. G., and Simon, H.A. (1958). *Organizations*. John Wiley and Sons, New York.

Müller, H.J., Malsch, T., and Schulz-Schaeffer, I. (1998). Socionics: Introduction and Potential. In: *Journal of Artificial Societies and Social Simulation* 1(3), <http://jasss.soc.surrey.ac.uk/JASSS/1/3/5.html>.

Parunak, H. Van Dyke (1996). Applications of Distributed Artificial Intelligence in Industry. In: O'Hare, G.M.P. and Jennings, N.R. (eds.): *Foundations of Distributed Artificial Intelligence*, John Wiley and Sons, New York, 139-164.

Perrow, C. (1986). *Complex Organizations: A Critical Essay*, third edition, Random House, New York.

Rammert, W. (1998). Giddens und die Gesellschaft der Heinzelmännchen. In: Malsch, T. (ed.): *Sozionik: Soziologische Ansichten über künstliche Sozialität*, Edition Sigma, Berlin, 91-128.

Schulz-Schaeffer, I., and Lührs, R. (1998). Sozialvorstellungen in der Verteilten Künstlichen Intelligenz, in: *Proceedings zum Workshop Sozionik auf der 22 Jahrestagung Künstliche Intelligenz*, Bremen, 15-17 September, 14-26.

Schulz-Schaeffer, I. (1998). Akteure, Aktanten und Agenten. Konstruktive und rekonstruktive Bemühungen um die Handlungsfähigkeit von Technik. In: Malsch, T. (ed.): *Sozionik: Soziologische Ansichten über künstliche Sozialität*, Edition Sigma, Berlin, 128-167.

Schulz-Schaeffer, I. (2000). Akteur-Netzwerk-Theorie. Zur Koevolution von Gesellschaft, Natur und Technik. In: Weyer, J. (ed.): *Soziale Netzwerke: Konzepte und Methoden der sozialwissenschaftlichen Netzwerkforschung*. R. Oldenbourg Verlag, München, 187-209.

Strübing, J. (1998). Multiagenten-Systeme als „Going Concern": Zur Zusammenarbeit von Informatik und Interaktionismus auf dem Gebiet der Verteilten Künstlichen Intelligenz. In: Malsch, T. (ed.): *Sozionik. Soziologische Ansichten über künstliche Sozialität*. Edition Sigma, Berlin, 59-89.

Thompson, J.D. (1967). *Organizations in Action: Social Science Bases of Administrative Theory*. McGraw-Hill, New York.

Wooldridge, M.J., and Jennings, N.R. (1995). Agent Theories, Architectures, and Languages: A Survey, in: Wooldridge, M.J. and Jennings, N.R. (eds.). *Intelligent Agents*, ECAI-94 Workshop on Agent Theories, Architectures and Languages, Amsterdam, 8-9 August 1994. Springer-Verlag, Berlin, 1-39.

HYBRID LEARNING COMMUNITIES IN AGENT-BASED INTELLIGENT DISTRIBUTED LEARNING ENVIRONMENTS

An Application for Socionics Research

ANDREAS HARRER
Fakultät für Informatik
Technische Universität München
Munich, Germany

Abstract

Intelligent Distributed Learning Environments (IDLE) are applications in which hybrid communities with human and artificial actors evolve. These learning systems are therefore an interesting research field for the new interdisciplinary area of socionics. We present the design and features of Intelligent Distributed Learning Environments and the actors that may participate in hybrid learning communities. The learning environment is designed as a multi-agent system with co-operative intelligent agents and human learners that may take different roles in the learning process.

1. Introduction

Intelligent Distributed Learning Environments (IDLE) are systems influenced by a number of research fields and different scientific disciplines. From the point of view of computer science they are at the crossroads between educational systems, artificial intelligence and (distributed) group work systems. There is also a strong influence from cognitive science, psychology, communication science and the social sciences. This influence arises because these systems provide an enviroment in which interactions between human and artificial participants take place and learning processes materialise. In this section we will present an overall picture of the decisive forces in the development of IDLE (see also *figure* 1).

1.1. INTELLIGENT TUTORING SYSTEMS

Intelligent Tutoring Systems (ITS) are a special type of program for Computer Aided Instruction (CAI). In general the computer can have different functions in

N.J. Saam and B. Schmidt (eds.)
Cooperative Agents.Applications in the Social Sciences, 165-182

CAI systems:

- The computer as **media**: A human teacher uses the computer as a means for supporting instruction. That may involve the use of simulations to explain some topic or the incorporation of modern hardware such as "lifeboards" which make possible the direct conversion of writing into computer formats like HTML.
- The computer as **information resource**: Here the computer is used to make additional sources of information available. Research using CDROM encyclopedias or through the World Wide Web can play a substantial role in the user's knowledge acquisition.
- The computer as **carrier of instruction**: The computer carries through planning and instruction itself. This may be in addition to conventional instruction or as a stand alone format for courses.

In Intelligent Tutoring Systems the computer is used primarily as the carrier of instruction.[1] The term **Intelligent** Tutoring System is used because ITS require competences which give it a certain amount of artificial intelligence:

- **domain competence**: The ability to solve problems in the domain of the ITS. A representation of the domain knowledge is needed for this.
- **diagnostic competence**: The ability to make an analysis of the correctness of the student's solutions and show how the current solution can be improved. In addition, the system needs to assess the current motivational state of the student. This competence is needed to develop a proper model of the student and thus make adaption to the individual learner's traits possible.
- **didactic competence**: The ability to adapt the learning process to the individual learner (their knowledge and motivational state). The system has to choose an instructional strategy appropriate for each learner.
- **communicative competence**: The ability to provide hints, corrections and other assistance in a way suited to the learner and offer the student a user interface they can use easily.

With all these competences an ITS is able to adapt its behaviour in the learning process according to the needs and traits of individual students. Based on the information collected in the student model, a proper instructional strategy is chosen which provides help to the student if necessary. It can explain correct solutions to the student and is even able to correct and complete the student's

[1] There may also be an additional function with the computer serving as an **information resource**. ITS often provide a hypertext document containing the contents of the course from which research can be carried out.

own solutions.

There are a lot of dependencies and influences between ITS and other research fields (see *figure 1*). **Cognitive science** and **artificial intelligence** are the main drivers for the implementation of domain competence. To make it possible for the computer to solve problems in the specific course domain, it is necessary to represent cognitive models and domain knowledge. At the same time, the use of cognitive models in ITS has had an impact on theories in cognitive science and provided a lot of interesting results for further investigation. **Motivational psychology** is fundamental for diagnosis of motivational states and traits of the learner and in order to adapt to the learner's motivation. The didactic competence is based on theories from **instructional science** and the implementation of these theories has made it possible to evaluate them using information collected in field studies with ITS. The design of the user interface for ITS is a concern influenced by the field of human computer interaction (HCI) based in **communication science**.

1.2. COMPUTER SUPPORTED COLLABORATIVE LEARNING

In recent years the widespread use of computer networks has facilitated the development of systems in which group work and collaborative learning processes are possible. Computer Supported Collaborative Learning (CSCL) systems provide support for interactions in learning groups. However, they do this only on the technical side (by means of shared and private workspaces, different channels of communication, e.g. chat systems similar to Internet Relay Chat, IRC). In these CSCL systems there has been no intelligent support of group processes similar to the support ITS offers for single users. The main forces influencing this approach are **distributed systems** in computer networks and the design of user interfaces which is the concern of HCI in **communication science**. The social interactions and phenomena[2] evolving in these virtual spaces where no direct face–to–face contact exists are also of particular interest to **techno–social science**.

1.3. INTELLIGENT DISTRIBUTED LEARNING ENVIRONMENTS

Intelligent Distributed Learning Environments (IDLE) bring together research in ITS and CSCL. The result is a learning system that facilitates group learning and supports the resulting learning communities using the kind of intelligent adaptation displayed by ITS. In *figure 1* we give a graphical summary of the forces and scientific disciplines important in the development of Intelligent Distributed Learning Environments.

[2] For example, a problem of group awareness (Bürger 1998) arises in distributed virtual rooms and a distributed environment has to take this into account if it is to support the group members appropriately.

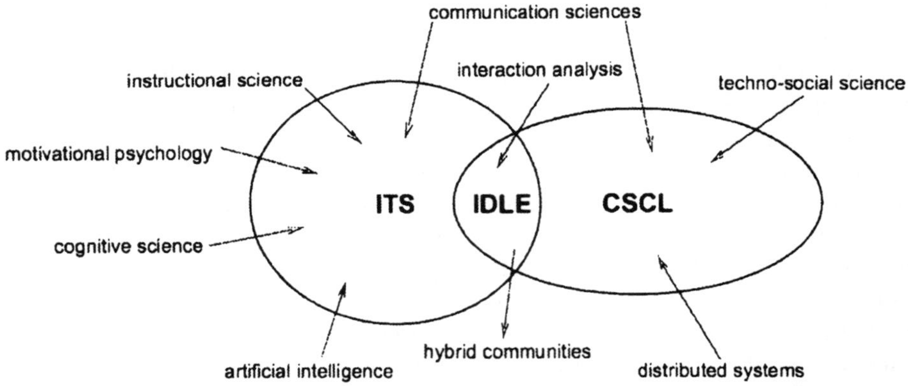

Figure 1: Influences on Intelligent Distributed Learning Environments

An IDLE supports learning communities in a number of ways. For example, it can play a role in group formation, composing groups well suited to learn together using the contents of the learner models as a criterion. It can also provide support using intelligent agents which have specific functions (take specific roles) in learning communities. To realise these IDLE features, we need two things. On one hand, we need mechanisms for **analysis and modelling of group interactions** to gain information representing the social aspects of the students' models. On the other, the artificial actors (agents) in learning communities need social competence to display behaviour that is appropriate for the role they are supposed to play. Since an IDLE is an environment for indirect interaction between participants and the communities in IDLE contain human as well as artificial actors (resulting in so–called **hybrid communities**) they also constitute an interesting topic of research for techno–social science. In the next section we will present the different kinds of actors in Intelligent Distributed Learning Environments and the interactions occuring in these hybrid communities.

2. Interactions in Hybrid Communities

In Intelligent Distributed Learning Environments we create a virtual space in which humans do not only interact with each other but also with artificial actors. These communities with human and artificial participants are called hybrid communities. In learning communities there are a number of typical patterns of behaviour and also a number of expected functions for the group which lead to specific **roles**.

2.1. ACTORS IN HYBRID LEARNING COMMUNITIES

Intelligent Distributed Learning Environments create an interaction space where human actors (the learners) and artificial actors (intelligent agents) exist. The interaction between these actors can be divided in three categories:

- **human–human** interaction: Since the learners are spatially or temporally separated, the software system has to provide the mechanisms to exchange messages between the users. In distributed systems there are often restrictions on the format and variety of these messages. Sometimes there is only a channel for text although additional audio or video channels may exist as well. The function of the software system as a messenger between human users is often called "computer–mediated communication" (CMC). In subsection 2.2.2. we will describe the design of interfaces usable for human–human and human–agent interactions in IDLE.
- **human–agent** interaction: To create "believable" agents which are accepted by the human users of a learning system, the artificial agents have to show behaviour that is useful for the learning community as a whole. It must also be appropriate for the function the agent is supposed to serve. For these ends, an agent must "understand" the interaction processes (at least partially) and must also have the ability to communicate with other actors in the community. In subsection 2.2.2. we will present our approach to interaction analysis which gives our agents a (limited) understanding of the flow of interaction processes and the contents of conversation. A similar technique can be used to produce messages meant for the agent's interaction with human actors during group learning.
- **agent–agent** interaction: This type of interaction in IDLE is only important for the internal processes of software systems and ultimately for their output behavior. Since this article concentrates on human interactions within IDLE and their consequences for research in social science, we will not discuss agent-agent interactions in detail. Artificial agents often communicate in standardised agent communication languages. Examples are FIPA–ACL[3] (FIPA 1998) or KQML[4] (Labrou and Finin 1997). They may also use pre-defined communication protocols for communication (for example the contract net protocol).

The actors in learning communities typically have different functions and therefore take special roles. The roles frequently occuring in these learning

---

[3] The Agent Communication Language designed by the Foundation for Intelligent Physical Agents.

[4] Knowledge Query and Manipulation Language, a well known agent communication language.

communities will be presented in the next section.

2.1.1. Roles in Learning Communities

In many types of ITS and CSCL systems, there were implicit assumptions about the functions which participants should play in the learning process. This involved an expectation of specific behaviour from the learners which is characteristic of roles (Schäfers 1986). To classify existing educational (ITS and CSCL) systems and provide a flexible means for describing different types of learning communities, we have defined a number of typical role profiles (Harrer 1999). Using this classification we will be able to integrate the other types of educational systems by combining these roles and assigning them to (human or artificial) participants in the learning community. Our role profiles are:

- **student**: This is the basic role in learning systems. Students want to expand their knowledge and also to improve their competences. In distributed systems that may well include improvement in the abilities needed to interact with other participants. A specific (but possibly faulty) part of the domain knowledge is necessary for this role.
- **tutor**: In situations where someone is needed to explain aspects of knowledge, the role of tutor comes into play. In most learning communities there will a tutor as well as the students. A tutor has the opportunity to enagage in teaching, planning and carrying through instructional processes. Additionally it is possible that the tutor's own domain knowledge will be expanded during the teaching process because reflection about the topic, which is necessary for explanation, can give rise to new insights. This technique is called "learning by teaching".
- **diagnostician**: If a learning community wants to have feedback about the current state of knowledge or the progress of a task solution, the role of diagnostician may be a sensible supplement. It is very well suited to learners needing a lot of feedback and to cases where the learner will benefit from an explicit diagnosis of their current state. The diagnostician has the opportunity to observe problem solving strategies and mistakes both of which could be unfamiliar or novel. This technique is called "learning by diagnosis". In addition, the diagnostician's ability to assess and evaluate actions and problem solving behaviour may improve while taking that role.
- **observer**: Information about the process and structure of group interactions can be valuable for the participants, especially in bigger groups. However, this information should be as unbiased as possible. Therefore in (non computer supported) collaborative and co-operative environments one or more observers are often present. Their task is to make an analysis and evaluation of the observed group processes. When social and communicative competences need to be improved, the results

produced by the observers are of particular benefit. The observer must have the ability to analyse, interpret and evaluate interaction between the participants in the learning community. A special class of observer is the "involved observer" who is not only observing but also taking part actively in group work. Such an observer will take on some aspects of other roles but this may produce biased observations because the observer cannot focus exclusively on the observing task.

- **mediator**: During group processes, situations often occur in which the regulation of communication and interaction may be useful. This may occur if there is a disagreement or the group reaches an impasse which it cannot resolve. In these situations, an intervention in the interaction process can provide a solution to this undesirable situation. The mediator's task is the detection of such situations and intervention to overcome them. The main abilities of this role are knowledge about regulation of interaction processes and the communicative abilities to perform interventions that help to break the deadlock in the current situation.

Existing types of educational systems often make implicit assumptions about the roles users and software components should play. The roles of student, tutor and diagnostician can be extracted from these systems. The roles of observer and mediator have not been implemented as autonomous agents in any existing system. They are also unsupported for human actors in these roles. It seems to us that all these roles are important for a flexible and encompassing variety of learning situations. Therefore our approach will support all of these role profiles, both for human actors (with appropriately adapted user interfaces which support playing these roles) and for artificial agents. In the next section, we give examples of the flexible composition of different learning situations with the help of our role templates.

2.1.2. Learning Situations Supported in IDLE

Our aim was to provide a flexible learning environment with a variety of different learning situations which is thus best suited to the different preferences and needs of users. By means of the role profiles and a combination of human and artificial actors taking these roles we can generate a great diversity of learning situations. In this way, we can integrate all the situations from existing types of educational systems. For example, the learning community known from classical ITS consists of one human student and one artificial tutor. In the variation described in Ritter (1997) we have learning systems with multiple tutors that may have different types or representations of domain knowledge and may communicate, co-operate or even compete with each other to help the student in the learning process. *Figure 2* shows these two learning situations in the graphical notation UML.[5]

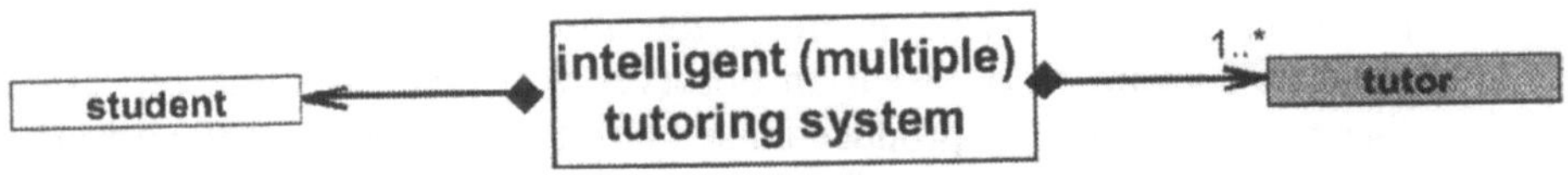

Figure 2: Learning situation: intelligent (multiple) tutoring system

The first systems that used intelligent agents for instruction were the "learning companion systems" (Chan and Baskin 1988) in which a human student has an artificial learning companion, i.e. an intelligent agent as a co-student. Additionally, both students can consult the (computer) tutor. The generalisation of this learning situation with the potential for multiple companions is shown in *figure 3*.

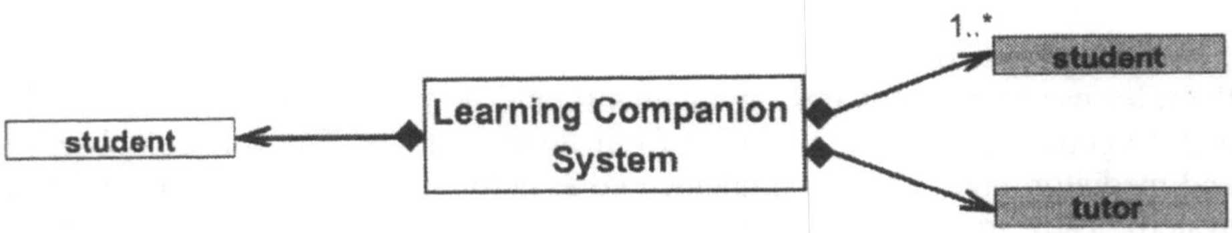

Figure 3: Learning situation: learning with artificial companion(s)

In Harrer (2000b) we showed in detail how existing types of educational systems could be integrated in our Intelligent Distributed Learning Environment.

If we add the roles of observer and mediator and consider all possible combinations of role profiles, there are now a great variety of different learning situations. Many of these are not available in existing ITS and CSCL systems. An example of a complex learning situation with human and artificial participants in various roles is presented in *figure 4*.

[5] Unified Modeling Language is the de facto standard for graphical notation in object-oriented software development. In our diagrams, the box in the centre represents the learning situation and the roles to the left are played by human actors. The roles on the right (shaded boxes) are taken by artificial agents. The number of actors in one specific role is given in the usual notation of "lower bound .. upper bound".

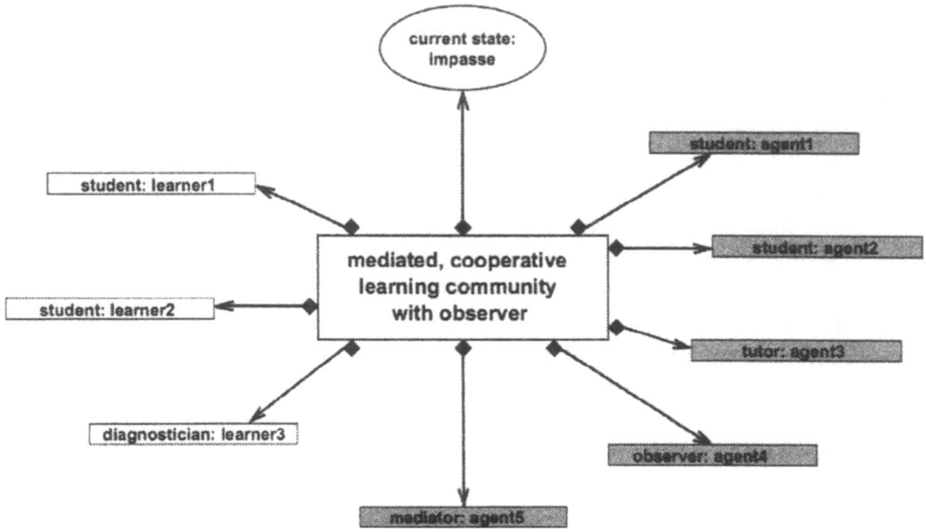

Figure 4: Learning situation: complex learning community

The learning situation is a co-operative[6] one because there is some division of labour between students and a diagnostician. An artificial observer provides the learning community with analysis and interpretation of their interaction behaviour. Since the current state of affairs is an impasse either the agent mediator has to intervene to regulate the communication process or the agent tutor will have to give hints to overcome it.

In the next subsection we will sketch our approach to the analysis of interactions which is necessary to make a diagnosis such as the one in the previous example: "the group is currently in an impasse".

2.2. ANALYSIS OF INTERACTIONS IN IDLE

Before we present an approach for analysis of group interactions in learning communities, we have to think about the sources of data that are available for analysis. The participants of learning communities can act at two different levels:

- **actions on the domain level**: These are taken to solve the tasks in the knowledge domain of the IDLE. User actions of this kind are handled by the domain level editors. In distributed environments an additional distinction can often be made between actions in private workspaces and actions in shared (public) workspaces. Actions at the domain level are of

[6] We use the term "co-operative" for situations with a division of labour and individual responsibilty for the completion of sub tasks and the term "collaborative" for a shared effort without such partitioning, following Burton and Brna (1996).

most use for the diagnosis of knowledge state, because they are tightly connected to the user's problem solving strategies.

- **interactions on the coordination level**: These are often characteristic for IDLE because in single user ITS there was no need for a coordination level. Participants use the coordination level to make plans for problem solving strategies on the domain level. The actions to solve the problem still happen on the domain level. On the coordination level the group members can carry out a number of tasks. They can make decisions about the division of labour between the participants. They can commit themselves to a specific role or function in the community and they can negotiate about the next steps to be taken on the domain level to make progress towards the learning goal. These social interactions are the primary source of data for the type of analysis we need in IDLE to get information about the current group state, structure and the meaning of interactions in the learning community.

Both sources of information should be investigated to collect data about the group processes in learning communities as comprehensively as possible. The reason why this information is so important for intelligent support of learning communities will be discussed in the following section.

2.2.1. The need for Student Modeling and Interaction Analysis

As we mentioned before a proper student model is crucial in the adaptation of an IDLE to the preferences and needs of the users in the current learning situation. In order to give hints with appropriate accuracy, choose a suitable instructional strategy and propose a group composition that is appropriate to the learner, we need cognitive, motivational, social and even physical information about each learner. One technique frequently used for complex information sources like student models is to separate the whole data set into different views which can be combined to provide a big picture. Therefore we partition our student models into the **individual learner model** and the **group model** as proposed in Paiva (1997). For an even finer distinction, we separated both these sub models into **cognitive**, **motivational**, **social** and **physical** aspects, a classification closely related to the one used in the agent reference model PECS (Urban 2000).

In addition to the uses of the student model data typical for ITS, there are other possible uses for intelligent support in IDLE:

- On one hand, we get the opportunity to support the formation of learning communities based on our student model data. This is particularly true in virtual spaces where learners do not know each other personally. Here the student model data is valuable for proposing the composition of groups with participants who are well matched according to their student model data. This approach was first discussed as "opportunistic group formation" (Ikeda, Go and Mizoguchi 1997). However, since there was no analysis of social interactions, the groups were built on the basis of cognitive aspects alone. We think that social aspects of the learner (like role preferences and level of participation) are important for proper composition of learning communities.

- On the other hand, we want to support the group work itself. To do this, the actions of intelligent role agents must be appropriate for the current situation as well as for the learners' individual abilities and preferences. For example in the learning situation shown in *figure 4* the artifical mediator agent could initiate a discussion to overcome the impasse. *Figure 5* shows a behaviour description for an artificial mediator mediating a discussion.[7]

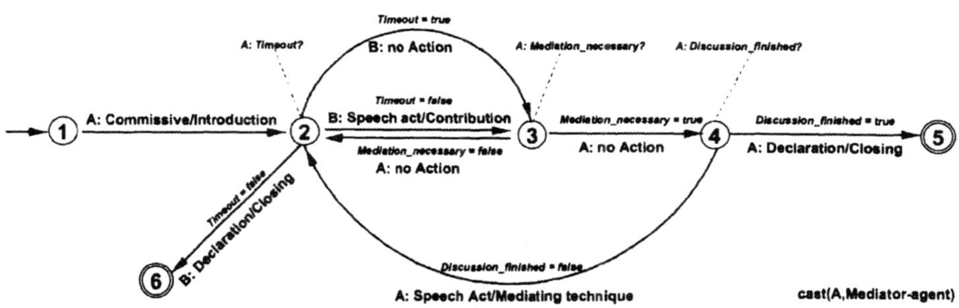

Figure 5: Behaviour description for mediator agent: mediated discussion

Transitions in the behaviour description are made according to the evaluation of predicates like `discussion_finished` which are computed by mechanisms for analysis of interactions. To make support by autonomous intelligent agents possible we need a rich student model and also a mechanism for the diagnosis of current events during group work.

Both of these support facilities are therefore heavily dependent on the ability of the system to carry out proper student modelling and analysis of interactions in learning communities. In the next section we will sketch our approach to this kind of analysis.

---

[7] Details of the notation and processing of such descriptions can be found in Harrer (2000b). The notation is based on the idea of conversational networks which will be discussed in the next subsection.

2.2.2. An approach for Interaction Analysis in Learning Communities

In our approach to interaction analysis, we base our methods on speech act theory (Searle 1997) and conversational theory (Winograd and Flores 1986). In this way, we have tried not to use an *ad hoc* approach but to develop one based on established theories and evaluation procedures.

Speech act theory is concerned with the analysis and classification of syntactic, semantic and pragmatic aspects of utterances. A speech act is a linguistic act that is intended to have some effect (on the hearer, the speaker or perhaps bystanders) when uttered. In contrast to statements, speech acts cannot be evaluated as "true" or "false". In speech act theory as presented by Searle (1997), five categories of speech acts are distinguished:

- Assertives: the speaker determines that something is valid
- Directives: the speaker wants to induce the hearer to do something
- Commisives: the speaker commits himself to an action
- Expressives: a psychic state is made explicit
- Declaratives: performing this speech act leads to conformity between words and reality

Each **performative** (utterance with which the speaker wants to cause an effect) can be assigned to one of these five speech act categories.

Conversational theory investigates typical patterns and sequences of speech acts in conversations. We analysed commonly occuring conversations taking place in an IDLE. To that end we extended the classification of speech acts and introduced sub categories to make finer distinctions possible. (For example, both a request and a command are directives, but their effect on a conversation may be quite different.) The observed patterns seem to be significantly domain independent and are therefore usable in IDLEs with different domain knowledge. Typically, these conversations are presented as conversational networks such as the one shown as a finite automaton in *figure 6*. (The end states of the conversation are shown in double circles and speech acts cause state transitions.)

This conversational network is a model of a conversation in which one actor requests an explanation (initial speech act from state *1* to state *2*) and another actor reacts to that request.

In Harrer (2000a) we extended the basic conversational network approach, intended for two person conversations, to conversational networks usable for group conversations. We introduced an additional layer, the so–called group network, to represent information about the learning group as a whole.[8] Our

[8] In this approach, we follow the principle that the whole is more than sum of its parts, which is related to the concept of "emergence" in the social sciences.

extended conversational networks make it possible to trace the flow of interaction in learning groups and to make a diagnosis of the current state a learning community is in. The prerequisite for this is to have a diagnosis of speech acts made by participants in the learning community. We will now present our approach for this kind of analysis.

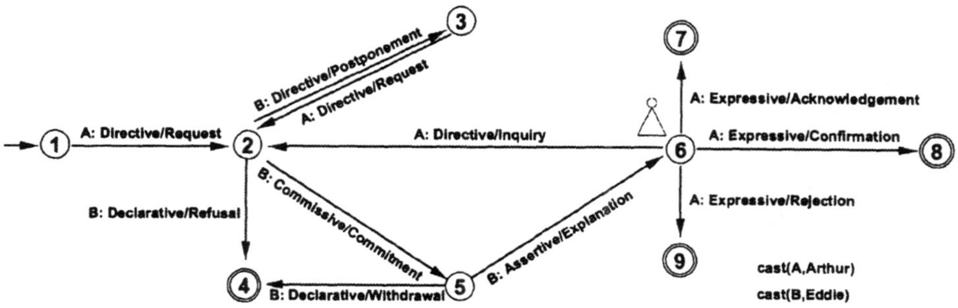

Figure 6: Conversational Network

The potential for the automated analysis of messages is often classified according to the degree of structure which a message has (Borghoff and Schlichter 1995). There are three broad categories:

- **unstructured messages**: The message is in a free format which is interpretable by human readers. For automated intepretation, sophisticated techniques of natural language processing would be necessary. A personal note without any predefined structure is an example of this message type.
- **semi–structured messages**: Some parts of the message are in a mandatory format but other parts are not predefined. The predefined parts can be interpreted and analysed automatically. The others can be seen as unstructured. An example of a semi structured message is an electronic mail. Specific parts, such as sender and receiver information, are in a fixed format but the content is unstructured and probably not automatically interpretable.
- **structured messages**: The whole content of the message is in a fixed format and therefore it can be interpreted completely by a computer. An example of this is a form for equipment orders with order numbers and quantities of each article.

When we design our interfaces for IDLE we have to ensure that human–human interaction is comfortable but also that human–agent interaction is capable of being analysed by the artificial agents. Otherwise they are prevented from providing intelligent support. Because of this difficulty, unstructured messages are not suitable for our aims. On the other hand, structured messages impose a lot of cognitive overhead on human users. They will have to become familiar with the format of the formal message language. This may be quite difficult, especially for users not acquainted with computer science. Such a structured interface would dramatically decrease the acceptance of the learning system. Instead, we decided to design our interfaces to produce semi structured messages. Human users are thus provided with a comfortable means for communicating with other actors and artificial agents can (at least partially) analyse and interpret the messages too.

Now we will present the design of our semi structured interface and give a detailed example for the use and interpretation of the messages it produces. For our semi structured user interface, we devised three different kinds of building blocks for messages, some interpretable by agents and some not:

- **predefined phrases**: These have specific information assigned to them which can be used for learner modelling and also interpreted by the IDLE.
- **choices**: The user can modify and parameterise the contents of these, for example by choice in a menu. These elements also have some machine interpretable information in them.
- **free user input**: This provides the full potential for users to express themselves. These parts are considered unstructured for the rest of this paper and therefore cannot be interpreted automatically without sophisticated use of natural language processing.

In *figure 7* we present an example of a sentence template in our semi structured interface:

```
Could you $Style explain to me $Intensity $Knowledge-unit
again? I'm not sure about the following: $Explanation
```

Figure 7: Example for a sentence template in the interface

This sentence template shows all three types of building block we can have in a semi structured message. The parts for `$Style`, `$Intensity` and `$Knowledge-unit` are choices for the user, by which they can choose a more polite form, a degree of urgency ("...explain to me now...") and a domain object (knowledge–unit) they want to refer to. The slot with `$Explanation` is free user input where the user can give details about the topic they want explained. The rest of the message is a predefined phrase, to which we assign a category which can be used to construct learner models. In *figure 8* we give an example of the way in which speech act information is assigned to the predefined phrase to give the message a machine interpretable meaning when the user chooses it.

```
Could you ...explain to me ...again?
I'm not sure about the following: ...
Category:     Directive
Subcategory:  Inquiry
Context:      refers to previous explanation
```

Figure 8: Predefined phrase with speech act information

We will now present a detailed example of the kind of information about the interactions in learning communities which our semi structured interface can produce. We assume a learning community consisting of two learners `Arthur` and `Eddie` who are taking part in a conversation represented by the conversational network in *figure 6*. The conversation is currently in state 6, actor *A* is `Arthur` and actor *B* is `Eddie`. The previous explanation `e1` didn't help actor *A* to understand the problem, so *A* chooses the message template in *figure 7* and produces the following message with the help of the semi structured interface:

> "Could you please explain to me now the initialisation of semaphore limit again? I'm not sure about the following: I don't know how to determine the correct value for the initialisation."

The IDLE uses the speech act information for the predefined phrase shown in *figure 8* to classify the kind of speech act as Directive/Inquiry. It then traces the state transition in the conversational network (from 6 to 2) which this speech act causes. Finally, it constructs an entry for the group model with information about the interaction that happened as a result of that speech act. This information unit is shown in *figure 9*.

```
Speech-Act:           i1
Sender:               Arthur
Receiver:             Eddie
Category:             Directive
Subcategory:          Inquiry
Intensity:            high
Position S-R:         equal
Context:              refers to Explanation e1
Propositional Content:
  (interpreted)       Operation_on_Semaphore(Initialisation,limit)
  (uninterpreted)     "I don't know how to determine the correct
                       value for the initialisation."
Style:                polite
```

Figure 9: An example for group model content created within the interface

Some of the relevant information is gained directly by the use of the predefined phrase (category or sub category of the speech act). Other information is indicated by the choices ("please" for a polite style, "now" for a high intensity) and by the conversational network (reference to explanation `e1`). There is also some propositional content in the message that can be interpreted by the system (the choice for `$Knowledge-unit`) and some that is not interpreted. This is the free user input part of the message. We can also make further inferences about that information entry that could provide hints about the roles the two participants have in the conversation. The interactions in the network could imply that `Arthur` is the student and `Eddie` is the teacher in this learning community.

In this example we show how messages are produced with the help of a semi structured interface and how information about the message can be extracted automatically for further processing by intelligent agents. This information enables the IDLE to trace the flow of interaction in conversational networks and to make a diagnosis of the current state of group work. The intelligent agents react to that information according to behaviour descriptions like that in *figure 5* in order to support the learning process. For a detailed description of agent behaviour and the support facilities in Intelligent Distributed Learning Environments refer to Harrer (2000b).

3. Social science research in intelligent distributed learning systems

Intelligent Distributed Learning Environments are an interesting area of research for the social sciences for several reasons:

- Distributed environments (IDLEs and the World Wide Web) give rise to a new type of interaction. This is not based on direct social contact but on contact at a distance that can be spatial or temporal (like asynchronous communication with email). It is not clear if established theories about direct interaction can be extended to those virtual distributed spaces or if new theories will have to be developed. One approach to the evaluation of virtual learning communities was developed in collaboration with the Institute of Sociology at the University of Munich and is presented in detail in Wagner (2000).
- In IDLE the socio–technical phenomenon of hybrid communities with human and artificial participants materialises. Under what circumstances do human actors collaborate with artificial agents and will these agents be accepted as entities or just be seen as tools? These question are important ones for the acceptance of IDLE by users and in the future for all types of interactive computer systems.
- The acceptance of artificial intelligent agents will surely be improved if

theories about the behaviour of actors from social science (such as rational choice) are implemented in artificial agents. The social sciences thus provide input to IDLEs that will have an effect on further studies of the emerging phenomena.

- If the agent's behaviour is developed using these theories it will then be possible to validate the theories using IDLEs as a "testbed": a special enviroment which is used for standardised test and comparison. An evaluation of the different theories in a given environment thus becomes feasible.

In the research discipline of socionics, the interests of social and computer scientists come together to investigate the processes by which new technologies – like computer networks – change human interaction behaviour. They also collaborate to build software systems that support human users as far as possible in achieving their aims. For IDLE that aim is obviously learning.

4. Conclusion

In this article we introduced a new type of educational system, the Intelligent Distributed Learning Environment (IDLE). We provided an overall picture of the influences on the development of IDLEs and their precursors. These learning systems provide an environment in which human actors (the learners) and artificial agents interact. The social space resulting from this is called a hybrid community. By combining typical role templates for learning situations we can arrange a great variety of learning situations in a flexible way. The artificial agents support the learning process by analysing interactions taking place in the learning community and adapting their behaviour to the current situation and the human users' needs. We sketched our approach for the automated analysis of interaction in IDLEs with the help of a special kind of semi structured user interface.

For the social sciences these systems provide an interesting field of research in virtual spaces and hybrid communities. At the same time, the performance and acceptance of these systems can be improved by implementing theories about interactional behaviour from the social sciences. Intelligent Distributed Learning Environments are therefore a field where the social and computer sciences can benefit from each other and fruitful co-operation is possible.

References

Borghoff, U.M. and Schlichter J.H. (1995). *Rechnergestützte Gruppenarbeit*. Springer-Verlag, Berlin.

Bürger, M. (1998). *Unterstützung von Awareness bei der Gruppenarbeit mit gemeinsamen Arbeitsbereichen*. PhD thesis, Technische Universität München.

Burton, M. and Brna, P. (1996). Clarissa: An exploration of collaboration through agent-based dialogue games. In: Brna, P., Paiva, A. and Self, J. (eds.). *Proceedings of the European Conference on Artificial Intelligence in Education EuroAIED'96*, Lisbon.

Chan T.-W. and Baskin, A.B. (1988). Studying with the prince: The computer as a learning companion. In: *Proceedings of the International Conference on Intelligent Tutoring Systems*, Montreal, 194-200.

FIPA (1998) Agent communication language. FIPA 97 Specification, Version 2.0 Part 2, Foundation for Intelligent Physical Agents, Geneva, Switzerland, 1998.

Harrer, A. (1999). Group learning in hybrid communities in intelligent tutoring systems. In: Lajoie, S.P. and Vivet, M. (eds.). *Proceedings of the World Conference on Artificial Intelligence in Education AI–ED 99*, IOS Press, Amsterdam, 708-710.

Harrer, A. (2000a). Interaction analysis at coordination level with semi–structured interfaces. In ECAI Workshop Notes, editor, *Analysis and Modelling of Collaborative Learning Interactions*, Humboldt University, Berlin, 17-22.

Harrer, A. (2000b). *Unterstützung von Lerngemeinschaften in verteilten intellgenten Lehrsystemen*. PhD thesis, Institut für Informatik, Technische Universität München.

Ikeda, M., Go, S. and Mizoguchi, R. (1997). Opportunistic group formation. In: du Boulay, B. and Mizoguchi, R. (eds.) *Proceedings of the World Conference on Artificial Intelligence in Education AI–ED 97*, Kobe, 167-174.

Labrou, Y. and Finin, T. (1997). A proposal for a new KQML specification. CSEE Technical Report CS-97-03, Department of Computer Science and Electrical Engineering, University of Maryland, MD, August.

Paiva, A. (1997). Learner modelling for collaborative learning environments. In: du Boulay, B. and Mizoguchi, R. (eds.). *Proceedings of the World Conference on Artificial Intelligence in Education AI–ED 97*, Kobe, 215-222.

Ritter, S. (1997). Communication, cooperation and competition among multiple tutor agents. In: du Boulay, B. and Mizoguchi, R. (eds.). *Proceedings of the World Conference on Artificial Intelligence in Education AI–ED 97*, Kobe, 31-38.

Schäfers, B. (ed.) (1986). *Grundbegriffe der Soziologie*. UTB Leske, Opladen.

Searle, J.R. (1997). *Expression and Meaning*. Cambridge University Press, Cambridge.

Urban, C. (2000). PECS: A reference model for the simulation of multi-agent systems. In: Suleiman, R., Troitzsch, K.G. and Gilbert, N. (eds.). *Tools and Techniques for Social Science Simulation*, Physica-Verlag, Heidelberg, 83-114.

Wagner, M. (2000). *Die Evaluation hybrider Lerngemeinschaften durch künstliche Agenten*. Masters thesis, Institut für Soziologie, Ludwigs–Maximilian–Universität München, May.

Winograd, T. and Flores, F. (1986). *Understanding Computers and Cognition*. Ablex Publishing Corporation, Norwood, NJ.

AGENT-BASED MODELING AND SIMULATION FOR HOSPITAL MANAGEMENT

RAINER SIBBEL
Chair of Production Planning and Industrial Management
University of Bayreuth
Bayreuth, Germany

and

CHRISTOPH URBAN
Chair of Operations Research and System Theory
University of Passau
Passau, Germany

Abstract

Traditional modelling approaches (mainly stemming from technically oriented application domains) have shown themselves increasingly inadequate in supporting the economic and organisational planning processes of hospitals. One of the major reasons for this problem is neglect of human decision making and behaviour as a relevant influence on the performance of such systems. In this paper, we show how the problem can be overcome by integrating agent technology into current modelling approaches. Simulation models of autonomous agents make it possible to take into consideration (and answer) questions about economic evaluation and process-oriented control more plausibly than current approaches are able to do.

1. Managerial Framework for German Hospitals

Hospitals have been an interesting object of research in industrial management for years (Helmig and Tscheulin 1998). On one hand, the health care system in Germany is of a growing economic interest that should not be underestimated. On the other, hospitals can be seen as providers of a typical service to which the special problems of service management apply. Finally, the legal and economic conditions for hospitals have been subject to steady change for a number of years.

Currently, there is an annual turnover of over 500 billion DM in German health care. This is much higher than in the so-called "key industries": the motor industry, chemicals and so on. In contrast to other industries, the apparent increase in health care turnover has been treated with scepticism. This is

N.J. Saam and B. Schmidt (eds.)
Cooperative Agents. Applications in the Social Sciences, 183-202

because the services provided are financed mainly by a compulsory insurance system which places great strains on the economy – especially since such systems tend to cause an over-use of resources (Heike 1994).

Special problems arise because health care has been growing faster than the national economy for the last few decades. Commonly, this is called the "cost explosion" and can be deduced from the steadily increasing premium rates of statutory health insurance.

In 1998, the costs of statutory health care in Germany amounted to over 97 billion DM. The absolute share of hospitals in the whole health care market is not incommensurately high. Therefore, hospitals are classified as the "cost drivers" because the hospital sector has had the highest rate of growth within the health care system during recent years.

Until some years ago, there was not much competition between hospitals – particularly because of statutory regulations. Since about 1993, the situation has changed considerably. Today, displacement competition between hospitals dominates the market in many regions. This competition is considered politically desirable and will intensify further in the future.

In addition, German hospitals are under pressure to adjust to a changing environment: medical and technological progress, demographic development and financial constraints. As a result, they not only have to face more intense competition but also have to cope with and co-ordinate more complex performance structures. Beside costs, with increasing customer orientation, quality turns out to be the determining factor for success. Integrated effectiveness and efficiency management in a hospital has to combine quality with economic goals.

Due to the substantial personnel component of performance in service provision, hospital employees are an influential factor (and offer the largest potential for success) in the quality of hospital performance and consequent patient satisfaction. A better customer orientation is necessarily connected with a process-oriented view and effective control of processes in hospitals. Particularly in service provision, effectiveness and efficiency depend directly on the choice of a performance program, the available capacities, the organisation of the necessary performance processes and the relevant managerial skills.

It can be seen that the requirements for hospital management have become more and more demanding in recent years and that this trend will continue to intensify in the future. Hospitals are on their way to developing into modern service companies. Because of these changes, it is necessary to develop adequate management tools to support the hospital administration in developing effective control.

2. Economic Fields of Application and Benefits of Agent-Based Simulation Models for Hospital Management

The centre of performance in a hospital is the extensive process of treatment, beginning with diagnosis and therapeutic measures, and ending with discharge or readmission to outpatient treatment (carried out by doctors with their own practices or other cost centres). Therefore, hospitals are instances of typical service enterprises that produce a very personal, highly complex and intangible outcome performed on and with the external factor 'patient'.

These special service characteristics lead to the conclusion that essential managerial decisions must be made both in the field of long-term planning for performance capabilities as well as in operational planning and control of existing capacities and processes (Schlüchtermann and Sibbel 1999). Strategic and tactical planning jointly determine the overall performance capability and its pattern over time. The fundamental economic problem is the fact that (together with the offered performance level) the majority of the resulting costs are being determined in advance.

The cost structure of person related and technology-intensive services (particularly characteristic of hospitals) is substantially influenced by decisions concerning potential demand. Apart from human resources, investment planning for operational resources (such as medical equipment and beds) is of central importance. Extraordinarily high fixed cost intensity forces management to co-ordinate the supply and demand of capacities. Therefore, it is desirable to achieve the highest possible flexibility in performance capability.

While capacity planning determines the general setting for processes, the scheduling and control of patients is also necessary for the time co-ordination of procedures in the hospital. Process level problems (like personnel placement, scheduling and sequencing) have a special meaning in service enterprises because economic performance depends crucially on the efficiency of process execution and the rate of use for available capabilities. In contrast to other industries where the paradigm of process orientation has already led to increased efficiency, hospitals are just beginning to participate in this development. Nevertheless, patient oriented process organisation is a necessary foundation for a customer-oriented strategy. For this reason, decentralised models of planning are at the centre of research interest.

Questions about logistics or workflow are often discussed from the perspective of non-economic goal criteria. However, this is not possible for practical problems of patient planning and control. By analogy with production planning and control, time objectives, such as lead-times, length of stay and time liability (or plant operating rates) are to be used as substitutes for economic goals. On the level of capacity planning, the evaluation of alternative decisions from the economic perspective is crucial because these decisions not only determine the majority of costs but are also difficult to reverse.

Due to the special nature of their tasks, hospitals are relatively overhead and fixed cost intensive. Therefore, the basis for the economic evaluation of alternative capacity decisions is the concept of process oriented control. The activity based cost accounting method is especially appropriate for service companies and hence plays a central role.

In this domain, the goal of agent-based simulation models should be to depict specific circumstances, processes and human decision behaviour in a hospital as accurately as is necessary to gain economically valid results. Based on such models, it should be possible to analyse alternatives and deal with the various managerial problems of capacity planning and patient control. In this context, it is important to integrate the resource oriented approach into process oriented design.

Including alternative cost accounting methods in such simulation models establishes the prerequisites for economic evaluation of alternative actions and for carrying out sound calculations and control computations in the context of decision oriented cost accounting. The emphasis of this approach is placed on cost unit accounting, based on process costing and cost centre accounting for profitability comparisons. The goal of integrating accounting approaches into such agent-based simulation models is to provide sound economic information for alternative decision situations in capacity planning and patient control.

3. Requirements for Simulation Models in the Hospital Domain

In the context of strategic and operative planning, management software systems are often used to deliver information for rational decisions in order to improve entrepreneurial performance. In particular, simulation models are of growing importance for investigating questions about capacity planning, organisation of processes and patient oriented scheduling. There are a number of requirements for such simulation models that will be discussed in the following sections.

The complexity of hospital procedures results from extensive performance bundles that involve numerous possible alternative processes. Performance related uncertainties make inclusive central planning and control all but impossible. Decentralised planning and control is necessary to take account of local conditions, competencies and responsibilities.

It needs to be borne in mind that both in the places where major decisions are made and where those decisions take effect, there are human beings involved and not rule based technocratic systems. Therefore, approaches that are restricted to modelling organisational structures, as they are used for production planning and control for example, are not useful. It is necessary to view the patients and medical personnel (nurses, medical-technical assistants and doctors) as agents who communicate and co-operate with each other. For this reason, an agent-based approach seems particularly plausible and promising.

If we are going to analyse alternatives in the planning and control of health services, we need to develop models that are better able to represent the significant circumstances of a hospital than the current ones. From the model results, we can derive useful decisions and meaningful planning and control strategies. This means an improvement in the structures and procedures of hospitals. Therefore, an agent-based planning system has to reflect the operating resources, the medical personnel, the patients, the organisation of procedures and the resulting costs. The individual decisions of the participants are of special importance. These individuals should be modelled as agents with their own autonomous behaviours and strategies that they can deploy to achieve their goals.

4. Modeling and Simulation in Hospital Management – State of the Art

Currently, simulation systems used in hospital management are based on the same concepts that are used in the simulation of manufacturing systems. Simulation models that represent organisational processes in hospitals consist of an arrangement of locations like wards or treatment rooms in which mobile units representing patients are treated on the basis of given therapy plans.

Facilities and resources are modelled by discrete-event components which can be integrated into a graphical layout describing the structure of the model. Later on, each component may be adapted to the real world situation by supplying adequate model parameters.

The medical staff and the doctors are modelled as passive resources. They are only given the status of facilities that must be present in order to carry out medical services for the patients. Local assignment and precedence strategies allocate the medical staff and doctors. The users of the simulation system may modify these allocations as they choose.

By analogy with work pieces in manufacturing systems, patients are often modelled as simple mobile components. They are equipped with standardised treatment plans, which describe the sequence of medical departments the patient has to visit over time. In most of the systems, it is assumed that these treatment plans are fixed. There is no possibility of dynamic adaptation to various phases of the therapy. An example of a tool that is currently applied to practical problems in hospital management and works in this way is given by the *Medmodel* system, developed by *Promodel Corporation* (Promodel 1999).

The major problem with using such simulation systems in the area of hospital management is the fact that human behaviour, which has a strong influence on the organisational processes in hospitals, is not taken sufficiently into consideration. On one hand, medical staff and doctors make decisions in given situations that are strongly based on the issue, the individual experience and the personality. These decisions change the processes in hospitals significantly and cannot therefore be modelled adequately by simple priority rules as used in available systems. On the other hand, it is not sufficient to model patients as passive work pieces that are controlled by static treatment plans. The individual properties of patients also play a much more important role than the one granted to them in current approaches to medical process modelling.

In contrast to the case of production planning and control, the possibility of forecasting and scheduling performance in a hospital is much smaller. At the same time, the influence of individual human decision-makers, the medical staff and the patients (as deliverers and recipients of diagnostic and therapeutic services) is much bigger. It is necessary for a planning system dealing with patient scheduling and control to consider the special features of the system, which result from uncertainty and human decision behaviour. For this reason, it seems particularly worthwhile to have a closer look at the agent-based modelling and simulation paradigm.

The major goal of this work is the development of an agent-based simulation system which will be able to support economic and organisational decision making processes in the hospital domain. In order to develop such a system we use a two step process model. In the first step, we intend to build a special purpose agent-based simulation model for a certain application context given by a real situation in a selected hospital. In this first step, we will analyse the basic requirements of simulation models in the hospital domain. Based on our experiences during this simulation study, we will try to generalise the concepts used as a second step and develop a general purpose, component oriented, agent-based simulation system that can be used as a general modelling framework in the hospital domain.

5. Step 1: Development of a Special-Purpose Simulation Model for a Real Situation in a Selected Hospital

5.1. REAL APPLICATION CONTEXT

In German hospitals, the balance between intensive and normal care units involves relatively few beds in intensive care. In cases of emergency (e.g. artificial respiration) all emergency equipment is also available in the normal care units. Experiences with a modified three or four level care system have shown that, with an alternative policy of capacity supply, it is possible to cut costs significantly without reducing the quality of care. The basic idea is to increase the capital-intensive capacity in intensive care (and the newly organised intermediate care units) in order to reduce the amount of equipment held in the normal care units.

For example, if patients on respirators stay longer in the intensive care units, then the numbers of respirators that are rarely used to full capacity in the normal care units can be reduced significantly. Besides, the risks for patients decrease in case of unexpected complications, because of intensive care and treatment regime. If the patients' care focuses more on each part of the treatment and the care units in which it can be done most effectively, than the patients can be more independent in the units before and after intensive care (in feeding for example). The increased costs caused by prolonging the stays of patients in intensive care are offset by higher use of capacity and significant reductions in costs for the units without intensive care.

These considerations will be illustrated using the example of a heart surgical ward. The central diagnostic and therapy process as it commonly occurs for most non-emergency patients (80% of the total) is shown in *figure 1*.

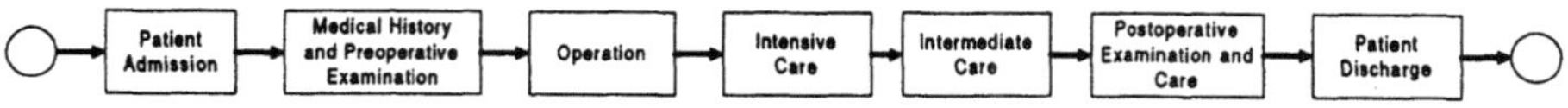

Figure 1: Systematic course of the treatment process in the heart surgical ward

Initially, the patient is asked to come into the hospital. First, administrative information is registered at the reception office. From there, members of the nursing staff take the patient to the assigned low care unit. The ward physician collects a medical history and carries out the first examination. This includes not only the standard preliminary measurements such as an electrocardiogram but also necessary preoperative processes such as X-ray analysis or microbiological tests. The ward physician also assigns the timetable for diagnosis and surgery and notes it in the medical record of the patient. Always attended by nursing staff, the patient passes through all the relevant functional units of the hospital. The order of the preliminary examinations is optional and decided by the nursing staff. The surgery is usually performed two or three days after admission.

After surgery, the patient remains in the wake-up unit for a short period of observation before being moved to the intensive care unit where s/he will be under permanent observation – usually for 24 hours. The patient is then moved to the intermediate care unit where s/he stays another 24 hours. The only difference between the two units is that the intermediate care unit is no longer equipped with respirators.

Subsequent care of the patient takes place in the normal care unit and lasts about seven days. During this time frequent inspections are made and re-examination and postoperative treatments are carried out when needed. The ward or senior physicians for each unit decide on the transfer of a patient from intensive to intermediate care and from intermediate to normal care. Finally, the patient is released from the heart surgical department for outpatient treatment.

Now, the aim is to develop an agent-based simulation model for the scenario described which should be able to depict the real conditions, procedures and human decision behaviour. It should also allow for an economic evaluation of different possibilities.

5.2. THE BASIC STEPS FOR THE DEVELOPMENT OF THE MODEL

This section describes the basic steps that are necessary to develop an agent-based simulation model in the hospital domain. The procedure is divided – like the general scientific cognitive process (Schmidt 1995) – into system analysis, model conception and realisation, model validation and execution of simulation experiments.

5.2.1. System Analysis

The system analysis stage is the one that focuses closely on the current circumstances in the hospital to be studied. This has to be done in the actual hospital itself. Besides the resources, such as rooms, machines and personnel, it is particularly important to record the individual process regulations and strategies that guide the decision makers. To produce valid inputs for the model, it is necessary to collect an adequate amount of data about the organisational processes and treated patients. This data must also be analysed carefully to classify symptoms as well as diagnostic and therapeutic measures. At this stage, the cost accounting methods should also be analysed and integrated into the model in order to enable an economic evaluation of the current situation.

The focus of the system analysis is the detail of the actual (real) system. The most important task at this stage is to identify and record the essential characteristics of the real system and separate them from the less important characteristics that are used later on for defining the limits of the system.

5.2.2. Model Conception and Model Implementation

Based on observation, an abstract model is developed, containing a formal description of the real system. The abstract model includes a set of model components with the relevant state variables and dynamic behaviour, the definition of the model structure and consideration of external influences at the boundary of the system. All three steps make use of the concepts of abstraction and idealisation. The following classes of model components seem to be of special interest here.

Resources. First of all, the facilities and resources needed for the treatment of the patients must be taken into account. Examples of such facilities are X-ray stations, laboratories, operation rooms, beds and wards. Generally, concepts for modelling these resources can be adopted wholesale from the simulation of manufacturing and production systems.

Human Factors. The most important aspect of this approach is the modelling of the human factor. This includes the medical personnel, the doctors and the patients.

As regards the medical personnel, knowledge and experience, operative planning strategies, communication and co-operation abilities seem to be of particular interest. The doctors are mainly characterised by their experience, status and responsibilities, and therefore influence the therapy plans of the patients. The patients are mainly "containers" for attributes affecting the organisational processes in the hospital. They have to carry information about their illnesses and therapy plans. The therapy plans describe the sequence of treatments that are necessary. These plans have to be open to changes over time. An example of such a therapy plan is given in *figure 2*.

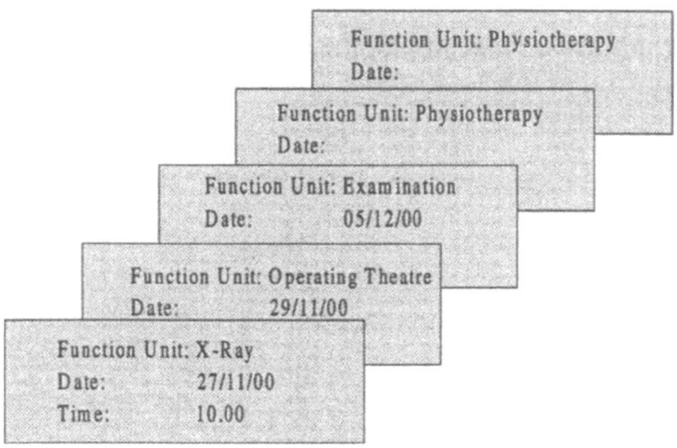

Figure 2: Example of a therapy plan

Communication Infrastructure. A communication infrastructure is also needed within the model. This enables the human decision-makers to exchange information in the context of their individual tasks.

Decisions and Strategies. In the planning and organisation of processes in a hospital, the problem arises that the overall goals of the hospital do not always coincide with the partial goals of individual functional units. If one chooses a completely decentralised mode of planning and leaves decision-making entirely to subordinate functional units, this will probably lead to an ideal mode of operation for partial goals. However, this mode of operation may mean that other functional units are hindered and restricted. This leads to unsatisfactory results for the overall goal that the hospital has set for itself. On the other hand, an exclusively centralised and global form of planning is bound to fail given the complexity of local processes and events in a hospital.

Therefore, our agent-based model should make a distinction between a global strategy and various local strategies. The global strategy sets approximate guidelines. These guidelines must be formulated in such a way that they do not excessively restrict the scope of the decision-makers operating according to local strategies. Nevertheless, they have to ensure that the overall goals the hospital has set for itself continue to be pursued. A similar dilemma occurs in the context of production planning and control. Promising approaches to these problems could first be tested for their applicability in hospitals and then be adopted in suitably modified form (Klinger 1999).

The specification of the decisions that have to be taken centrally and of the global strategy required in individual instances is a difficult problem. A simulation model provides a powerful tool here. It is capable of trying out numerous planning alternatives. It can then pinpoint the organisational forms that bear the global goals in mind while allowing local goals to be pursued at the same time.

The example in *figure 3* illustrates the places where decisions and strategies play an important role in our case study.

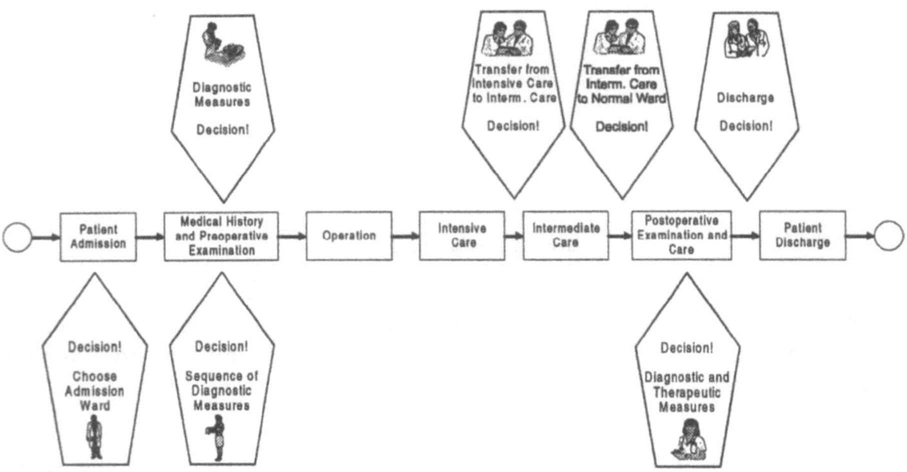

Figure 3: Decentralised decisions exemplified for the Heart Surgical Ward

While the patient is able to influence the outcome of the treatment process by his/her state of health (expressed in the duration of treatment and stay), the physicians decide on the necessary examination and treatment processes as well as the duration of stay in the different units. Finally, members of the nursing staff determine the order and exact temporal requirements for the examinations to be conducted on the patient. Usually, there are only limited resources of personnel and operating funds at the disposal of medical staff and their allocation is decided locally in each unit.

Model Implementation. Model implementation focuses on the transformation of the abstract model into an executable simulation. For this purpose, it is necessary to consider the set of concepts in the chosen model description language because these set up the formal framework of the simulation model.

The information collected during the system analysis stage forms the foundation for the model conception and realisation. Here, the most important challenge is to integrate agent technology and cost accounting into the simulation model. The essential model components – the machines and resources – have to be represented in the simulation. At the same time, the patients, doctors, and the nursing personnel have to be implemented as PECS agents. In this context, personality related attributes and characteristics, knowledge about the environment, and decision strategies as well as communication and negotiation abilities need to be taken into consideration. Depending on their symptoms, cyclical medical schedules must be assigned to patients. These contain sequences of medical and nursing services and must adjust dynamically to the service procedures and decisions of the people involved.

5.2.3. Model Validation

To prove the usefulness and capacities of the agent-based simulation model, the validation stage has to show that the model results correspond (within acceptable bounds) to circumstances in a real hospital. This requires the analysis of simulation data which have to be compared with the data found in the real system using suitable indicators. A successful validation is one of the major prerequisites for the practical adoption of such planning methods.

If the model does not approximate reality within satisfactory bounds, it has to be refined and extended accordingly. As soon as the agent-based simulation model has been validated, new conditions and alternative actions can be introduced into the model and investigated for their future consequences.

5.2.4. Experiments and Requirements for Simulation Results

As already stated above, the validated simulation model should be used for planning and forecasting purposes. In order to evaluate alternative organisational structures based on a simulation model, we must define criteria that characterise the quality of the simulation results. The following criteria seem to be useful for this purpose.

A first indicator is provided by capacities. For example, the numbers and operation times of human resources (doctors, medical staff) or the number of beds and quantities of medical equipment for each department could be used here. A second indicator is the workload of resources like beds and medical equipment. A third indicator is waiting times of patients for diagnostic or therapeutic treatments. All of these will provide useful information about the overall performance of the system.

In the context of customer orientation, criteria concerning the patients are of particular interest. For example, the distribution of times for different processes (examination, care, transport and recreation) or the resting times between processes should be available as simulation results.

In order to measure the performance of the health services offered, values for throughput times should be collected in various departments, as well as total numbers of patients served within a specified time horizon.

Last but not least, the resulting costs should be accessible. These costs are an important indicator of hospital performance. In particular, the simulation should indicate the total amount of generated income, the total costs, mean values of the costs for each patient and perhaps process costs.

5.3. APPLICATION OF THE PECS REFERENCE MODEL

The PECS (Physis, Emotion, Cognition and Social Status) reference model (Urban 2000) implements autonomous agents that are able to represent the complex behaviour of real actors. PECS assumes that, in order to model real individuals, it is necessary not only to represent rational and cognitive abilities but also to take physical, emotional, and social aspects into consideration.

The PECS architecture is based on a general, component oriented and hierarchical structure, which must be adapted to the requirements of particular application domains. The components provided by PECS cover the modelling of environmental influences, an infrastructure for information exchange between the agents and a set of components that constitute the internal state and the behaviour of the agents themselves.

In order to apply the PECS reference model to hospital management, a system analysis must first be carried out in the hospital. This analysis must gather information about the decision making and behaviour of the medical staff and the doctors, the interactions of the relevant actors and the organisational structure of the system.

Based on this data, we construct agents that act as representatives of the real world entities. The agents should be able to display the characteristics, behaviour and types of decisions of their real world counterparts. The major problem in this context is that the behaviour and decisions of the medical staff and doctors which are observed in reality must be simplified, abstracted, and idealised in order to reduce the agents' complexity. At the same time, the agents have to be kept complex enough that it is possible to transfer the simulation results to the real situation in a valid way.

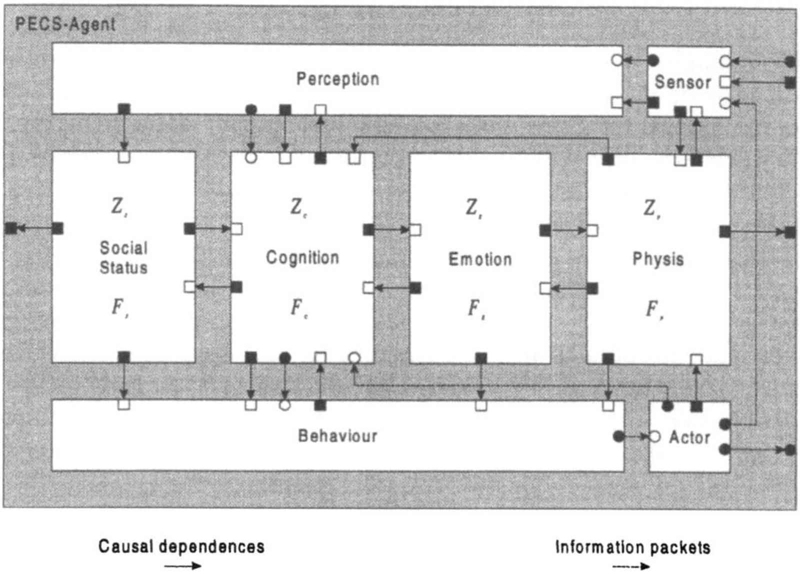

Figure4: Basic Architecture of PECS-Agents

The interactions observed between the actors in the real system may be administered by the PECS *Connector* component within the model. The *Connector* component provides a kind of central switchboard that enables direct message interchange between the agents as well as a blackboard for asynchronous communication. For example, negotiations carried out by members of the medical staff to schedule the treatment of patients can be done using the *Connector* mechanism.

The given organisational structure of the hospital, including its different departments or resources, defines the environment for the agents. For modelling these kinds of influences, basic concepts can be taken from the field of simulation in manufacturing mentioned above.

6. Step 2: Development of an Agent-Based Simulation System for the Hospital Domain

The major goal to be achieved in the future is the development of a general agent-based simulation system for the hospital domain. The experiences gathered during the simulation study that was described above will be useful to improve the quality and applicability of the resulting simulation system. The following section will give a brief introduction to the major design decisions for such a simulation system.

6.1. COMPONENT-ORIENTED MODELLING

The simulation system realised will be component-oriented. This means that we will provide pre-defined model components that enable the users to build simulation models for given real situations very quickly and effectively. Based on this model, the user will be able to perform simulation experiments in order to examine various alternatives concerning the organisation and its control structures which are relevant for the given situation.

The internal structure of the simulation system could be organised as shown in *figure 5*. The various functional units, (radiology, wards, laboratories and so on) are at the centre of the simulation system. A therapy plan is assigned to every patient and contains the place and time for each treatment or service like a checklist. On the basis of this therapy plan, the patient is led from function unit to function unit. The medical staff work on the basis of an appointment plan which contains the times and places of the services to be performed for various patients.

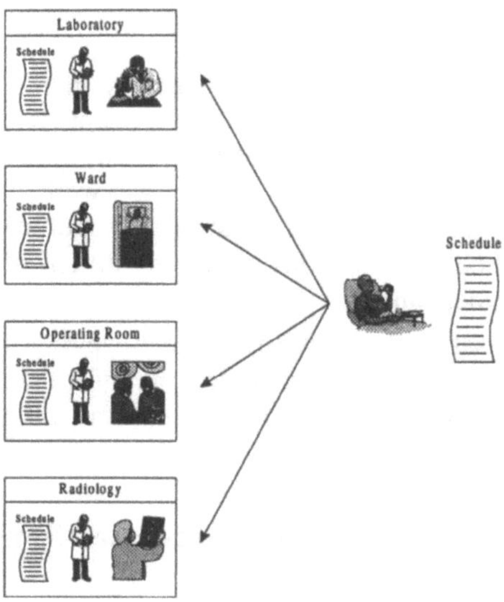

Figure 5: Internal structure of the simulation system

The planning problem now is the harmonisation of the patients' checklists with the appointment plans of the medical staff. As already noted in the previous chapters, global planning and the allocation of performance needs and performance offers is impossible given the complex situation in a hospital. Decentralised solutions have to be found. In particular, this means that neither the checklist nor the appointment plan can be completed in a single planning step. They are dynamically completed according to the circumstances and local decisions and strategies.

6.2. GRAPHICAL MODEL DESCRIPTION

The user of the simulation system will be provided with the option of describing individual models in a graphical way. For this purpose, the system will include a graphical model editor. This could work in the following way:

The left side of the work space offers a number of pre-defined model components like operating room, radiology, ward and laboratory. These components can be selected and dragged into the drawing area on the right hand side. The individual components are equipped with a number of specific ports that can be used to connect the components with each other. Each inserted component may be supplied with individual parameters in order to customise it for the real situation.

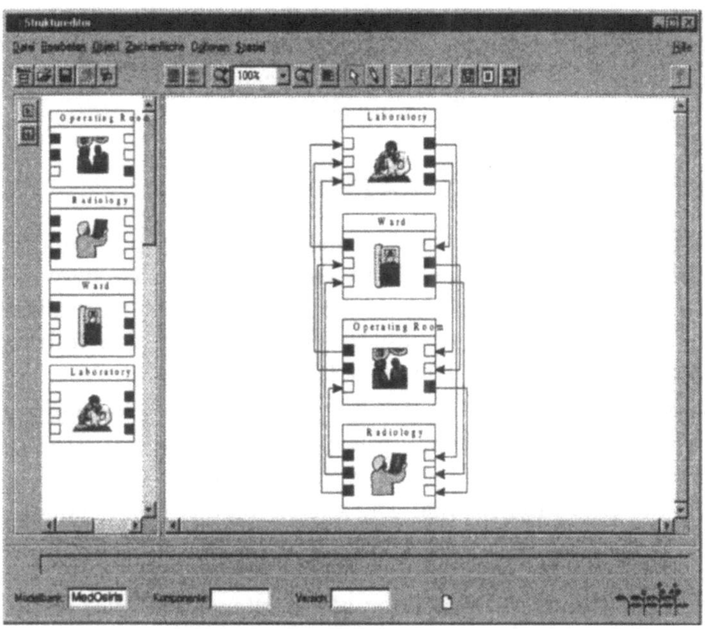

Figure 6: Screenshot of the Graphical Model Editor

In order to produce more clarity, it will also be possible to describe hierarchically structured models. This means that model components may consist of a set of sub-components on a lower level of abstraction. *Figure 7* shows an example of such a component. The component *Radiology* consists of eight interacting sub-components which model the internal structure and behaviour of a radiology department.

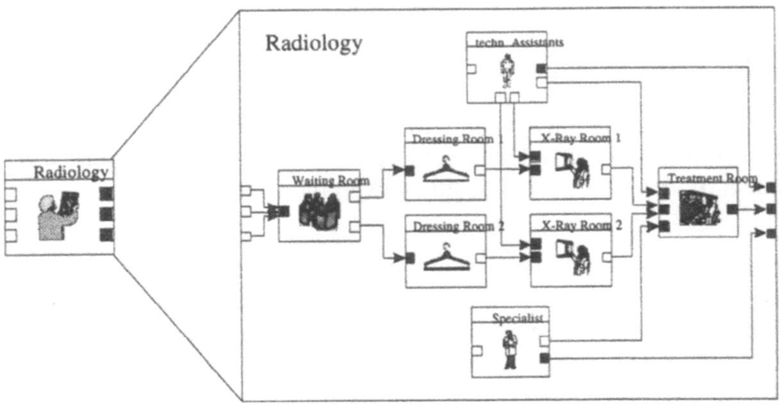

Figure 7: Example for a hierarchically structured model component

When the model definition is completed, the graphical representation will be transformed into an executable simulation program automatically. Graphical model construction is a technology that has already been successfully applied to the area of simulation in manufacturing and logistics.

6.3. SPECIFICATION OF STRATEGIES

As the local decisions of doctors and the medical staff are particularly important to our simulation system, it is necessary to provide a means by which the user can specify the decision behaviour of the agents used to represent them. It is intended to introduce a strategy description mechanism that follows the SSA reference model (Schmidt and Toussaint 1996).

In general, a strategy can be defined as a plan for a sequence of actions which should lead a system from a given initial state to a desired target state. According to SSA, a strategy consists of three major parts: the strategy call, the strategy and the execution section. The strategy call specifies conditions under which a given strategy has to be called. The strategy is composed of a set of rules which describe the execution instructions that have to be chosen under certain given conditions. A calculation block extends the rules and enables the introduction of algorithmic elements. After a certain decision has been made, the corresponding execution instruction is forwarded to the executor. The execution section administers the information about how the different actions have to be executed and also triggers the execution of those actions.

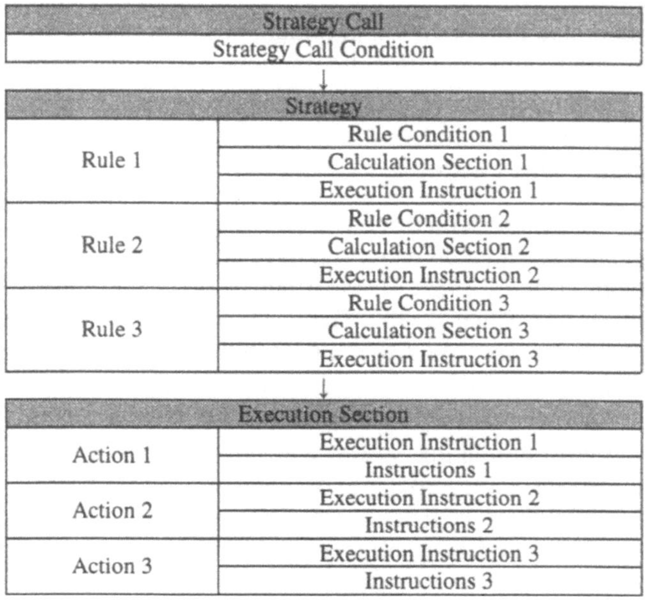

Figure 8: Basic structure of strategies according to the SSA reference model

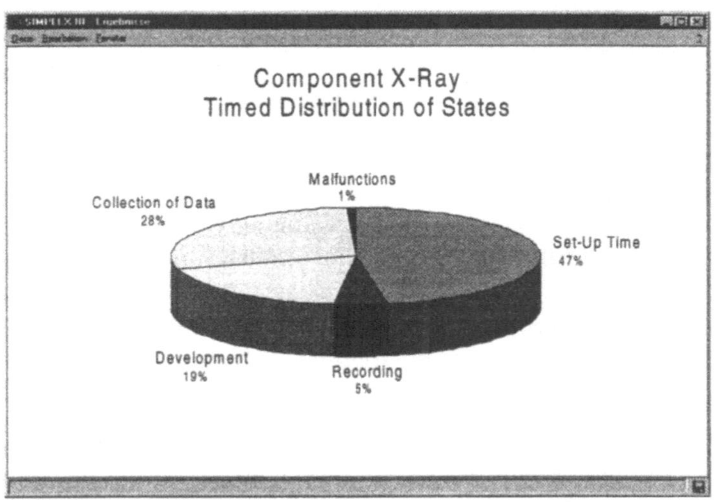

Figure 9: Time distribution of states for a given X-ray station

6.4. PRESENTATION OF SIMULATION RESULTS

For all model components, it will be possible to observe calculated data during the simulation runs. After a simulation run, the collected data can be analysed in more detail and also presented via graphical business charts. For example, the following pie chart shows the time distribution of possible states for a given X-ray station.

7. Summary and Outlook

Nowadays, component-oriented simulation systems belong to the state of the art in technologically oriented domains and have also found their way into the field of public health (Davies 1985, Harris 1986, Wright 1987, Davies and Davies 1995, Appelrath and Sauer 1998). The models that are currently used in the health care domain are usually dominated by the classical paradigms of technical simulation and the systems embedding those models usually display the advantages and characteristics of modern simulation systems.

However, these simulation systems generally neglect the very important impact of human decision making and behaviour on the performance of such systems. State-of-the-art systems are not agent-based. If the human factor is considered in those models at all, human beings are simply seen as passive work pieces or resources which have to be present in order to provide some kind of service. It is no surprise that, under these assumptions, current modelling approaches largely fail to provide adequate support for economic decisions in the hospital domain.

By integrating agent-based approaches into classical simulation systems, it is possible to construct simulation systems that can cope with influences stemming from the individual decisions and actions of human beings. In this way, the integration of agents into traditional simulation software will permit increased psychological realism in complex simulation models. This will obviously improve the quality of results that can be obtained by simulation models that have to deal with human influences in some sense. Current approaches to agent architectures, (such as the PECS reference model) seem to be very promising in this context and should, therefore, be integrated into such planning systems.

The specific properties of the medical sector and the nature of process performance in hospitals lead to the following conclusions:

- The simulation of processes and workflow in hospitals cannot be carried out on the basis of purely deterministic or stochastic approaches because the relevant effects of human (decision) behaviour are neglected or represented inadequately.
- The decentralised structure of the organisation, decisions within strongly personalised performance processes and the patient as an external human

factor all make an approach using agent-based simulation extremely useful. The people involved with medical services, their interests, their requests, their competencies, their individual properties and emotions and the combination of these aspects can be represented more realistically using agents.

- A simulation model which provides an authentic representation of the performance processes in a hospital serves as a basis for analysing the configuration and design of potential structures like the layout of stations or the capacity of beds and personnel.
- A simulation model using agents raises the possibility of taking into consideration (and answering) questions of economic evaluation and process-oriented control.

The goal of agent-based simulation models must be to depict concrete circumstances, processes and human decision behaviour in hospitals as authentically as possible and to support economic evaluations. Alternative actions for the different economic problems of capacity planning and patient scheduling and control can be analysed as well.

The crucial purpose of this approach is the integration of the resource-oriented view and the process oriented design of workflow with the help of the agent-based technology. The reason for including cost accounting approaches in the planning system is to generate economically relevant costing data which can be used for decisions in the field of capacity planning, patient scheduling and patient control. As another part of process oriented control, criteria and target ratios have to be chosen and modified so they can support goal-oriented management and take into account the tension between formal economic considerations and substantive medical goals.

References

Appelrath, H.-J., and Sauer, J. (1998). MEDICUS: ein medizinisches Ressourcenplanungssystem. In: *Künstliche Intelligenz*, 12: 56-60.

Davies, R.M. (1985). An interactive simulation in the health service. In: *Journal of Operational Research Society*, 36: 597-606.

Davies, H., and Davies, R. (1995). Simulating health systems: modelling problems and software solutions. In: *European Journal of Operational Research*, 87: 35-44.

Harris, R.A. (1986). Hospital bed requirements planning. In: *European Journal of Operational Research*, 25: 121-126.

Heike, A. (1994). *Problemgeschichte der Gesundheitsökonomik in der Bundesrepublik Deutschland. Die ökonomische Steuerung von Angebot und Nachfrage im Gesundheitswesen von der Kostenexplosion bis zum Gesundheitsstrukturgesetz.* Verlag Kohlhammer, Köln.

Helmig, B., and Tscheulin, D.K. (1998). Krankenhausmanagement in der deutschsprachigen betriebswirtschaftlichen Forschung im internationalen Vergleich. In: *Zeitschrift für Betriebswirtschaft (ZfB)*, 68: 83-110.

Klinger, A. (1999). *Referenzmodelle für die Abbildung von Personalsteuerung in der Simulation.* SCS, Erlangen.

Promodel: http://www.promodel.com/products/medmodel, 1999.

Schlüchtermann, J., and Sibbel, R. (1999). Produktionsplanung und -steuerung (PPS) für Dienstleistungsunternehmen. In: Corsten, H., and Schneider, H. (eds.): *Wettbewerbsfaktor Dienstleistung*. Verlag Vahlen, München, 59-77.

Schmidt, B. (1995). *Simplex II – Ein integriertes Simulationssystem – Benutzerhandbuch*. SCS, Erlangen.

Schmidt, B., and Toussaint, A. (1996). Das Referenzmodell SSA für Strategien. In: *SiP – Simulation in Passau*, Heft 3: 8-15.

Urban, C. (2000). PECS – A Reference Model for the Simulation of Multi-Agent Systems. In: Suleiman, R., Troitzsch, K.G., and Gilbert, G.N. (eds.): *Tools and Techniques for Social Science Simulation*. Physica-Verlag, Heidelberg, 83-114.

Wright, M.B. (1987). The application of a surgical bed simulation model. In: *European Journal of Operational Research*, 32: 26-32.

Author Index

Subject index

THEORY AND DECISION LIBRARY

SERIES A: PHILOSOPHY AND METHODOLOGY OF THE SOCIAL SCIENCES

Editors: W. Leinfellner (*Vienna*) and G. Eberlein (*Munich*)

1. G. Zecha and P. Weingartner (eds.): *Conscience: An Interdisciplinary View*. Salzburg Colloquium on Ethics in the Sciences and Humanities (1984). 1987 ISBN 90-277-2452-0
2. R.W. Scholz: *Cognitive Strategies in Stochastic Thinking*. 1987 ISBN 90-277-2454-7
3. H. Nurmi: *Comparing Voting Systems*. 1987 ISBN 90-277-2600-0
4. M. Schmid and F.M. Wuketits (eds.): *Evolutionary Theory in Social Science*. 1987 ISBN 90-277-2612-4
5. C.L. Sheng: *A New Approach to Utilitarianism*. A Unified Utilitarian Theory and Its Application to Distributive Justice. 1991 ISBN 0-7923-1301-1
6. F. Forman: *The Metaphysics of Liberty*. 1989 ISBN 0-7923-0080-7
7. G. Bernard: *Principia Economica*. 1989 ISBN 0-7923-0186-2
8. J.W. Sutherland: *Towards a Strategic Management and Decision Technology*. Modern Approaches to Organizational Planning and Positioning. 1989 ISBN 0-7923-0245-1
9. C. Vlek and G. Cvetkovich (eds.): *Social Decision Methodology for Technological Projects*. 1989 ISBN 0-7923-0371-7
10. P. Hoyningen-Huene and F.M. Wuketits (eds.): *Reductionism and Systems Theory in the Life Sciences*. Some Problems and Perspectives. 1989 ISBN 0-7923-0375-X
11. K.G. Grunert and F. Ölander (eds.): *Understanding Economic Behaviour*. 1989 ISBN 0-7923-0482-9
12. G. Antonides: *The Lifetime of a Durable Good*. An Economic Psychological Approach. 1990 ISBN 0-7923-0574-4
13. G.M. von Furstenberg (ed.): *Acting under Uncertainty*. Multidisciplinary Conceptions. 1990 ISBN 0-7923-9063-6
14. R.L. Dukes: *Worlds Apart*. Collective Action in Simulated Agrarian and Industrial Societies. 1990 ISBN 0-7923-0620-1
15. S.O. Funtowicz and J.R. Ravetz: *Uncertainty and Quality in Science for Policy*. 1990 ISBN 0-7923-0799-2
16. J. Götschl (ed.): *Erwin Schrödinger's World View*. The Dynamics of Knowledge and Reality. 1992 ISBN 0-7923-1694-0
17. R. von Schomberg (ed.): *Science, Politics and Morality*. Scientific Uncertainty and Decision Making. 1993 ISBN 0-7923-1997-4
18. J. Baron: *Morality and Rational Choice*. 1993 ISBN 0-7923-2276-2
19. M. Allais and O. Hagen (eds.): *Cardinalism*. A Fundamental Approach. 1994 ISBN 0-7923-2398-X
20. E. Carlson: *Consequentialism Reconsidered*. 1995 ISBN 0-7923-3547-3
21. J. Götschl (ed.): *Revolutionary Changes in Understanding Man and Society*. 1995 ISBN 0-7923-3627-5
22. A. Camacho: *Division of Labor, Variability, Coordination, and the Theory of Firms and Markets*. 1996 ISBN 0-7923-3832-4
23. R. Hegselmann, U. Mueller and K.G. Troitzsch (eds.): *Modelling and Simulation in the Social Sciences from the Philosophy of Science Point of View*. 1996 ISBN 0-7923-4125-2

24. J. Nida-Rümelin: *Economic Rationality and Practical Reason*. 1997 ISBN 0-7923-4493-6
25. G. Barbiroli: *The Dynamics of Technology*. A Methodological Framework for Techno-Economic Analyses. 1997 ISBN 0-7923-4756-0
26. O. Weinberger: *Alternative Action Theory*. Simultaneously a Critique of Georg Henrik von Wright's Practical Philosophy. 1998 ISBN 0-7923-5184-3
27. A. Carsetti (ed.): *Functional Models of Cognition*. Self-Organizing Dynamics and Semantic Structures in Cognitive Systems. 2000 ISBN 0-7923-6072-9
28. J. Nida-Rümelin and W. Spohn (eds.): *Rationality, Rules, and Structure*. 2000 ISBN 0-7923-6326-4
29. J. Klüver: *The Dynamics and Evolution of Social Systems*. New Foundations of a Mathematical Sociology. 2000 ISBN 0-7923-6443-0
30. V.A. Lefebvre: *Algebra of Conscience*. Revised Edition with a Second Part with a new Foreword by Anatol Rapoport. 2001 ISBN 0-7923-7121-6
31. J. Götschl (ed.): *Evolution and Progress in Democracies*. Towards New Foundations of a Knowledge Society. 2001 ISBN 1-4020-0063-4
32. N.J. Saam and B. Schmidt (eds.): *Cooperative Agents*. Applications in the Social Sciences. 2001 ISBN 1-4020-0190-8

MIX
Papier aus verantwortungsvollen Quellen
Paper from responsible sources
FSC® C105338

If you have any concerns about our products,
you can contact us on
ProductSafety@springernature.com

In case Publisher is established outside the EU,
the EU authorized representative is:
Springer Nature Customer Service Center GmbH
Europaplatz 3, 69115 Heidelberg, Germany

Printed by Libri Plureos GmbH
in Hamburg, Germany